HOW TO BE SECULAR

A Call to Arms for Religious Freedom

Jacques Berlinerblau

MARINER BOOKS
HOUGHTON MIFFLIN HARCOURT
BOSTON | NEW YORK

First Mariner Books edition 2013

www.hmhbooks.com

Library of Congress Cataloging-in-Publication Data
Berlinerblau, Jacques.
How to be secular : a call to arms for religious freedom / Jacques Berlinerblau.
p. cm.
Includes bibliographical references and index.
ISBN 978-0-547-47334-5 ISBN 978-0-544-10516-4 (pbk.)
1. Secularism — United States. 2. Freedom of religion — United States.
3. United States — Religion. 4. Church and state — United States. I. Title.
BL2747.8.B477 2012
211'.6 — dc23
2012014226

Book design by Brian Moore

Printed in the United States of America
DOC 10 9 8 7 6 5 4 3 2 1

Praise for

HOW TO BE SECULAR

"*How to Be Secular* serves as an important reminder . . . that we protect our rights to our personal beliefs by preserving the rights of our neighbors to believe otherwise. I agree wholeheartedly with Berlinerblau's argument and highly recommend this powerful book."

—Mario M. Cuomo, former New York State governor

"In this new look at church-state relations in America, Berlinerblau manages to be serious and sprightly in equal measure. This is a call to reject extremism of any sort and return to the American genius for accommodation of our differences—even, indeed especially, our differences over the role of religion in our public life."

—Elliot Abrams, former deputy national security advisor

"As someone whose faith is an important part of his life, I highly recommend this book and Berlinblau's defense of religious freedom. With great insight and clarity, he explains why it is important to protect and preserve secularism as a philosophy and he then lays out a twelve step program to revive it.

—Ambassador Dennis Ross, counselor to the Washington Institute for Near East Policy and former US peace envoy to the Middle East

"Berlinerblau succeeds in making concrete the current threats to secularism and offers a reasoned blueprint for an organized secular movement to regain its political power.

—*Publishers Weekly*

"An impassioned argument for 'a firm and dignified defense of the imperiled secularish virtues and moderation, toleration, and self-criticism.'"

—*Kirkus*

To the memory of Pasquale Spadavecchia (1931–2008)
Country Doctor. Philosopher. Gentleman.

"You had religious education?"

"None that you could take seriously."

"I pity you." So flatly stated that he might as well have been telling me the time.

"Yes, you feel sorry for me?"

"Secular don't know what they are living for."

"I can see how to you it might look that way."

"Secular are coming back. Jews worse than you."

"Really? How much worse?"

"I don't like to say even."

"What is it? Drugs? Sex? Money?"

"Worse. Come, mister. It'll be mitzvah, mister."

If I was correctly reading his persistence, my secularism represented to him nothing more than a slightly ridiculous mistake.

– PHILIP ROTH, *The Counterlife*

Contents

Preface

A FEW DAYS BEFORE the commemoration at Ground Zero marking the tenth anniversary of the World Trade Center tragedy, an article appeared on the front page of the *New York Times* titled "Omitting Clergy from 9/11 Ceremony Prompts Protest."[1] The protesters in question were incensed over Mayor Michael Bloomberg's decision to prohibit religious officials from speaking at the memorial honoring those who perished in the attacks.

Uproar aside, the decision was entirely consistent with the city's usual practice. In the past, services on this day of national mourning did not typically feature official representatives of religious groups. That clerics were not invited to participate on September 11, 2011, was neither unprecedented nor unusual. But leaders of the Christian Right suddenly deemed this arrangement unacceptable.

Richard Land, a major figure in the ultraconservative and highly influential Southern Baptist Convention, was quoted in the article as saying, "We're not France . . . Mr. Bloomberg is pretending we're a secular society, and we are not."[2] Elsewhere, Land lamented that Bloomberg's action "demonstrates the mindless secularist prejudice of the political establishment on our nation's Eastern Seaboard."[3]

Doing its own reporting on the growing controversy, the *Christian Century* cited the president of the Institute on Religion and Democracy, who charged that New York's mayor was "ignoring most Americans and most New Yorkers by pretending religion is unimportant."[4] The same article mentioned a right-wing blogger who accused Bloomberg of launching a "de facto jihad" on religion.[5]

As the controversy crescendoed, the mayor's detractors certainly had reason to believe that they might prevail. After all, in recent decades the Christian Right had been routing American secularism (a term that, as we shall see, has been defined, derided, used, and abused in a bewildering variety of ways). The growing influence of this movement was evident in the manner in which faith and piety had come to permeate the rhetoric of politicians and, ultimately, law and policy. Reproductive rights had been checked across the country—so much so that legal abortions are extraordinarily difficult to procure in many states of the Union. Science's role in shaping national dialogue on questions such as the teaching of evolution or the threat of climate change had been degraded. American public education had been challenged by attempts to de-secularize the curriculum or even remove students from its institutions via voucher programs or homeschooling.

Now the conservative Christian "outrage machine" was revving and whirring again.[6] And in New York City, for the love of God! This must have seemed like a heaven-sent opportunity for the assembled activists. If the Christian Right could make it unsecular there, they could make it unsecular anywhere! Imagine the mayor of the most secular city in America, and possibly the world, being forced to bend to the will of a few Bible-thumpin' pastors from the boonies!

That did not come to pass. In the face of a brutal battering from the media, Bloomberg held his ground, often with truculence. The fact that he had always maintained cordial relations with the city's diverse communities of faith certainly strengthened his position.[7] Another explanation for his triumph was the refusal of the Catholic Church to join the evangelical Protestants who had initiated the scrum.[8] The ceremony proceeded solemnly and without incident. The critics quietly decamped from this theater of the culture wars. In all likelihood they'll be back for another go.

This book is about the recent crackup of American secularism and the therapeutic steps required for its rehabilitation. In understanding how the patient became institutionalized (or, more precisely, deinstitutionalized) we will need to make sense of its complex historical past. In order to secure the future of secularism we will need to understand what secularism is and, more important, *what it is not.* According to its enemies, secularism is akin to atheism, hatred of religion, anti-Americanism, and—why not?—radical jihadism.

The stakes are very high. Were secularism to completely collapse, the country might become the type of "Christian nation" that the New York protesters hope (and pray) for. In the mid-twentieth century, the U.S. Supreme Court assiduously labored to purge any such possibility from our political system. Yet the form of secularism they abided by has fallen upon some very difficult times. New ideas, new energy, and most of all new people are needed to resurrect it in America today.

To a large extent, the ceremony of September 11, 2011, exemplified the possibilities of a new vision for secular America, wherein both freedom *of* and freedom *from* religion are granted as much space as possible. Under the Bloomberg protocols, no state-sanctioned clergy or prayer was included during the memorial. Yet those citizens who wished to express their faith were completely at liberty to do so.

President Obama opened his remarks by quoting from the Psalms.[9] Rudy Giuliani read from the book of Ecclesiastes.[10] George W. Bush invoked Abraham Lincoln's letter of condolence to Mrs. Bixby, which ends with the words "I pray that our heavenly Father may assuage the anguish of your bereavement, and leave you only the cherished memory of the loved and lost, and the solemn pride that must be yours to have laid so costly a sacrifice upon the altar of freedom."[11]

Yet some of the speakers said not a word about God. Perhaps they refrained because they wished to keep their faith to themselves. Or perhaps they did not believe in God. One imagines that Bloomberg's no-clergy directive offered these mourners one less distraction on a day of sorrow.

Still others invoked religion with moderation, dignity, and restraint, which are hallmarks of the secular worldview. One mourner, Debra Epps, noted that her brother's name was imprinted on the 9/11 memorial next to that of another victim, Wayne Russo. The men had sat next to each other at work. The family of the latter had called and asked for permission for their names to be enshrined side by side on the monument. "Christopher would have loved knowing," Ms. Epps explained, as she closed a short speech otherwise light in faith-based themes, "that the love he freely gave to others was given back to us in his name. Thank you, and I bid you God's speed."[12]

Introduction: Is Secularism Dead?

> Secularism is the handy one-word distillation for all that is wrong in the modern world. Consumerism, divorce, drugs, Harry Potter, prostitution, Twitter, relativism, Big Brother, lack of moral compass, lack of community cohesion, lack of moral values, vajazzling – all can be lumped together and explained by the word *secular*, a kind of contemporary contraction of *heathen* and *barbarian*, with undertones of greed, perfidity, and vulgarity.
>
> – Caspar Melville, *"Mix and Match Secularism"*

American secularism is in a very bad way. Conservative religious leaders rampage against it, demagogues denounce it on the campaign trail, all three branches of government give it the cold shoulder, and among the general public it suffers from a distressing lack of popular appeal. All of these are worrisome developments. But in the triage ward currently housing the secular predicament, one illness demands our immediate attention: a debilitating confusion as to what secularism actually is.

The idea of secularism has been in play for centuries – some might say millennia.[1] At the assorted genius bars of Western civilization, it has long been one of the regulars, and as such it has been defined in a lot of plausible ways, many of which will be discussed in this book. Yet the following definition seems powerful, precise, and the most conducive to its survival: *Secularism is a political philosophy, which, at its core, is preoccupied with, and often deeply suspicious of, any and all relations between government and religion. It translates that preoccupation into various strategies of governance, all of which seek to balance two necessities: (1) the individual citizen's need for freedom of, or freedom from, religion, and (2) a state's need to maintain order.*

When secularism achieves that balance – and sometimes it fails disastrously – it performs a public service that should always evoke awe. It enables citizens to live peacefully as equals with other citizens whose creed is different from their own. Secularism, then, is a political philosophy about governance that can bestow a secondary "bonus" effect: it may create or actualize certain dispositions and worldviews in us all. Foremost among these are the "secularish" qualities, such as tolerance toward others, moderation, and a willingness to be self-critical about one's own faith.

As for its primary function, secularism guarantees that this country belongs as much to a Sikh American as it does to a member of the nation's Protestant majority. It ensures that your child is not forced to join a "voluntary" prayer circle in the school cafeteria. It renders the authorities powerless to regulate any aspect of the consensual sex you will have tonight (secularism, by the way, wishes you all the best on that joyous occasion).

If you have ever marveled at the comparative lack of interreligious strife in America, you might want to say "*Thank you, secularism!*" In fact, why haven't you said so already? After all, it has defended a reading of the Constitution's inscrutable Establishment Clause that has done you a monumental favor: keeping federal and state government from molding you into an obedient subject of someone else's religion. And if you fancy being able to think about God in any way you see fit, then once again, a little gratitude is in order. This type of freedom is secularism's essence. This is secularism's promise. This is the end to which all genuine secularisms aspire.

But this is not, to put it mildly, how critics in the United States and abroad see things. They don't associate secularism with peace, freedom, and order. Rather, they equate it with godlessness, totalitarianism, and genocidal regimes. Secularism is depicted as a pervasive moral evil capable of undermining entire societies, or perhaps civilization itself. Many critics view it as a cancer. Still others treat it as a corpse: secularism had its heyday and was fun while it lasted. But now it's gone – skip the funeral.

The turpitude of this concept has been depicted in many colorful ways. To get a sense of the range of this commentary, we start with Pope Benedict XVI. Not a fan of secularism, but an erudite commentator nonetheless, the pontiff has depicted it as a spiritual menace that leads us away from our "ultimate purpose" because it forces us to treat religion as a "private matter."[2] Elsewhere he cautions that "there is something deeply alien about the absolute secularism that is developing in the West . . . a world without God has no future."[3] Conservative Protestants in this country take a different tack. They have spent decades preaching that secularism is literally demonic.[4] One American Christian fundamentalist worried aloud that "today, within the bounds of the Church, we are witnessing a Satanic work of deception and substitution that is intended to deceive even the very elect."[5]

Politicians express the same concerns but in a more secular idiom, if you will. The former British prime minister Tony Blair called on all faiths to join together against secularism.[6] In December 2007, the U.S. presidential candidate Mitt Romney likened it to "radical Jihadism."[7] Newt Gingrich has often expressed his disdain for "secular fanatics trying to redesign America in their image."[8] He then proceeded to write a book decrying Barack Obama's "secular-socialist machine."[9]

This last accusation is quite ironic since President Obama himself has, in word and in deed, attacked secularism. We will get to the deeds later, but you might recall that in his book *The Audacity of Hope* he chided fellow Democrats for equating "tolerance with secularism."[10] By drawing this equation, argued Obama, his party had foolishly "forfeit[ed] the moral language that would help infuse our policies with a larger meaning."[11] The Democrats, incidentally, got the memo and promptly delivered a brisk stiff-arm to the secularists within their traditional base. Loss of old allies: another headache for secularism.

Some detractors have moved beyond the critique stage and are performing a deathwatch. A conservative think-tanker deems secularism "a view of life ill-equipped to meet the political and existential challenges of the twenty-first century."[12] Pastor Rick Warren, one of the most prominent evangelical leaders in the nation, counsels, "You need to understand that the future of the world is not secularism. It's religious pluralism. You may not like it, but I'm sorry, that's it. The world is becoming more religious, not less."[13]

For Warren, there is an inherent contradiction between religious pluralism and secularism. He seems unaware of the close and complementary relationship between these ideas. The very concept of secular government arose during the Reformation and Enlightenment in order to safeguard religious pluralism, which theocratic governments were congenitally incapable of ensuring. Warren fails to understand that secularism, far from being the enemy of religious pluralism, is its *guarantor*. With a similar lack of precision Warren assumes that one cannot be both religious and secular. This flawed idea is pervasive and, as we shall see, has prevented many potential advocates from wholeheartedly adopting the ideals of secularism.

The same misconception can be observed among those who have already proceeded to the eulogy stage, and they are not necessarily religious conservatives. One distinguished theologian of the Left writes that "secularism is dead . . . people are turning to religious explanations for human existence and human meaning, and turning away from reason, science, and materialism."[14] Another professor opines that "secularism has had a reasonably good life and has done some good to the society but has now exhausted its possibilities."[15]

These and other diagnoses as to the continuing viability of secularism are astonishing not only for the breadth and boldness of their conclusions but also for the ideological range of those who promulgate them. Those who criticize secularism are liberals and conservatives, Republicans and Democrats, Catholics and Protestants, professors and politicians. They batter away at the idea to a degree that seems gratuitous and cowardly. Giving secularism a good concussing is largely a risk-free undertaking because in public debate convincing counterarguments have not gained traction. Will anyone come to the defense of the secular virtues?

The Revival: Every Day Will Be Sunday

It is not a coincidence that these oncologists and eulogists of secularism render their verdicts in an era of religious reawakening.

For decades social scientists have been tracking the global resurgence of religion much in the way that scientists monitor a tornado from a chase van. Years ago sociologists such as Daniel Bell and Robert Wuthnow described this trend as the "return of the sacred," or the "rediscovery of the sacred."[16] A little while back another sociologist, Peter Berger, noted that "conservative or orthodox or traditionalist movements . . . are on the rise almost everywhere."[17] A more recent book title also gets the gist of the matter pretty well: *God Is Back: How the Global Revival of Faith Is Changing the World.*[18]

Let's refer to this phenomenon, as manifested in the United States and elsewhere, as "the Revival." It is the Revival that has lifted the spirits, loosened the lips, and steeled the resolve of the countless critics already mentioned. It is the Revival that threatens to turn American secularism – once a high-performance technology – into a quaint obsolete curiosity, like those first-generation iPods, which weighed more than a jar of mayonnaise. And it is the Revival that is perhaps irredeemably changing our world.

As it has taken shape from the 1970s forward, the Revival demonstrates two distinct modes: (1) lawful and somewhat alarming and (2) militant and terrifying. If visual aides are necessary to help conceptualize the latter then radical Islamists have provided us with a singularly horrific highlight reel. From al-Qaeda's weaponizing of passenger planes in downtown New York and elsewhere, to the "*Hey, let's film our martyrdom videos in high def!*" zealotry of Hamas and Hezbollah, to the blast-a-Buddha antics of the Taliban, images of jihadist violence are a reminder that the rise of religious expression has many troubling aspects.

Yet these frightening groups are only a marginal component of the phenomenon in question. When it comes to the Revival, militancy is the exception, not the rule. Wholly *lawful* actors have played the most important role in the dramatic political shifts that have disoriented the proponents of secularism over the past few decades.

Unlike their extremist counterparts, lawful Revivalists rarely resort to violence. They don't need to. They don't need to murder hostages or launch planes into buildings (whether they would do so under political circumstances that impeded their advance is an open question). All they require is the democratic structures, set in place by their secular enemies, to do their democratic thing and acknowledge the will of the majority. That's because Revivalists not only have God on their side, but usually the masses as well.

A recent study demonstrates startling demographic shifts that should keep secularists awake at night. Eric Kaufmann, author of *Shall the Religious Inherit the Earth?*, makes the case that "religious fundamentalists are on course to take over the world through demography."[19] Kaufmann notes that Muslim women in urban areas who are "most in favour of Sharia [law] bear twice as many children as Muslim women who are least in favour."[20] A secular couple in Tel Aviv in the 1990s produced 2.27 kids, while their ultra-orthodox counterparts brought forth a whopping 7.61 young ones.[21] In country after country the pattern is the same: religious traditionalists bring forth their basketfuls and quivers of progeny. Meanwhile, birth rates among more secular populations are stagnating or declining.

As for the United States, consider a statistic reported by Michael Lind: per every thousand women in Mormon Utah there are ninety children born every year; in more secular and liberal Vermont, there are forty-nine.[22] As the fruitful multiply more and more, Revivalists can pursue their agendas by working *legally* within the existing political structures of constitutional democracies.

Americans have experienced the lawful Revival through the forty-year ascent of the Christian Right. In this movement, white evangelical and fundamentalist Protestants strategically team up with traditionalist Catholics, conservative Mormons, and small groups of Orthodox Jews. The evangelicals, however, are the stars of this show. As often happens among lawful Revivalists, they display the virtues of effective leadership, financial discipline, institutional creativity, and organizational dynamism.

Many political scientists argue that the evangelicals formed the core of the "values voters" who swung the election to George W. Bush in 2004.[23] The forty-third president was presiding over an unpopular war and a sluggish economy. Yet he prevailed, in large part because he received the support of nearly 80 percent of the nation's evangelicals.[24] That's nothing to overlook, since evangelicals comprise one-quarter of the American electorate. When Obama, incidentally, sloughed off secularism he was simply making a pragmatic calculation, doing the electoral math. But he and the Democrats may have arrived too late: white conservative evangelicals, as anyone who has followed the GOP's drawn-out presidential primaries knows, have become a mainstay of the party.

It would be wrong to conclude that all evangelicals are politically conservative. Still, in its most organized manifestation the movement's intentions for America and secularism do not diverge much from the threat expressed by the Reverend Jerry Falwell, a fundamentalist, thirty years ago: "The godless minority of treacherous individuals who have been permitted to formulate national policy must now realize that they do not represent the majority. They must be made to see that moral Americans are a powerful group who will no longer permit them to destroy our country with their godless, liberal philosophies."[25]

Today, the "moral Americans" to whom Falwell alluded are complaining that a tyrannical secularism has recently pushed faith – not evangelical Christianity, mind you, but "faith" – out of public life. One wonders, though, whether evangelicals themselves actually believe this. The secularism they are excoriating now is precisely the same secularism that they have pummeled for decades at the state and federal levels. As we will observe, secularism – small, disorganized, underfunded, often represented by myopic and overheated ideologues – is the best oppressor conservative Christians ever had.

In any case, the fruits of secularism, as far as the Revivalists are concerned, are easily discernible and deeply disconcerting: schoolchildren are denied instruction about "faith" in public institutions of learning; science has dethroned theology as the "engine of civil policy."[26] Public officials are prevented from openly praying to God on behalf of our sinful nation.

According to the Revivalists, issues related to gender and sexuality have gone hopelessly awry as well. Women, they lament, have been permitted and even encouraged to reject their traditional childbearing role. Reproductive freedoms have been extended to the point that birth control and abortion are not just legal but normative. Pornography is readily accessible – on one's own personal computer, no less! The entertainment industry celebrates non-monogamous and non-heterosexual lifestyles. Gays and lesbians are becoming increasingly visible and accepted in society. The Revivalists have set their sights on undoing these social changes; they are making considerable headway.

Secularists who have taken note of this evangelical activism feel a sense of dread. They fear that this return of the sacred will transmogrify into the *reign* of the sacred. They fret that such lawful groups are only tactically and temporarily lawful. That is to say, once they achieve power they will dissolve the secular structures and safeguards that stand as the crowning glory of the American political experiment.

Some secularists imagine that the Talibanization of the United States might go down like this: On Monday, the Revivalists will move their operations into the public square, where the secular virtue of toleration will ensure that they will be accorded full rights of participation. On Tuesday, they will seize political power through their ability to amass huge blocs of voters (thereby annulling the sacred precept of the citizen who votes based on individual conscience). On Wednesday, they will collapse the distinction between public and private, unleashing squadrons of morals police to monitor speech, sex, art, thought, what have you. And then every day will be Sunday.

The Religious Moderates: The Future of Secularism

Undoubtedly this gloomy prediction raises some valid points. Yet it overlooks one formidable firewall against the Revivalist onslaught: religious moderates. Many self-described secularists today ignore this group, erroneously assuming, just like Pastor Warren, that religious people cannot espouse secular values. Many religious moderates, in turn, refuse to ally themselves with secularism, sometimes because its self-appointed spokespersons scare the bejesus out of them or just annoy them.

The role of religious moderates in checking the Revival has not received much attention. For one thing, journalists are not very interested in the subject. A more sensational front in the culture wars – the Manichean, thermonuclear free-for-all that constitutes the debate between Revivalists and extreme atheists – has absorbed the attention of the media. Such debates, and the cacophonous buzz they generate, draw focus away from a crucial truism: the decisive cultural conflict in this country is not between nonbelievers and believers; the real hot ideological action is occurring among members of the *same* faith. Revivalists abhor not only secularism, but also what they perceive to be its handmaiden: liberal theology and its moderate religious views.

At present a pitched ideological battle rages between mainline Protestants and conservative evangelical Protestants, between Left-leaning and conservative Catholics, between progressive and traditionalist Muslims, and between liberal and ultra-orthodox Jews. This is the conflict that will shape the future of this country, and it engages the entire slate of divisive national issues, ranging from abortion rights to gay rights to foreign policy.

For every group in the United States that resembles the Catholic League for Religious and Civil Rights (widely viewed as a right-wing advocacy organization), an antagonist like the liberal Catholics United has arisen. For every Muslim organization extolling the virtues of sharia law, there is a countermovement of Muslim dissidents trying to block its path. For every archconservative evangelical church with a massive Community Worship Center, there is a mainline church with a social justice agenda. Whereas Orthodox Jews condemn homosexuality as an abomination, Reform Judaism ordains gay clergy in impressive numbers.

In an odd and inadvertent way, the religious moderates – every bit as God-loving as their Revivalist counterparts – are doing the work of secularism. They are clearly on the defensive, but through church-by-church, mosque-by-mosque activism, they are impeding the advance of the lawful extremists. Many of these moderates are opposed to excessive mingling of government and religion. In this sense they are secular – *they are the future of secularism!* But getting them to see themselves as such is difficult.

Their reluctance to embrace this identity may rest on two incorrect assumptions they make about secularism (the problem of definitions again!). First, they assume it is the same thing as atheism, and not just any type of atheism. Secularism today is increasingly defended by the small confrontational group known as the New Atheists. Brimming with bravado, these polemicists insist that their numbers are swelling and their political power is burgeoning.[27] They point to statistics—which they misinterpret in a most self-serving way—stating that there are nearly 30 million Americans just like them. As the New Atheists see it, they are poised to put the fundies (by which they mean all religious people) in their place.

The result of having extreme anti-theists carry the secular flag has taken its toll. Secularism is now frequently associated not only with atheism but with radical anti-theism, or hatred of religion. The New Atheists, for their part, have decided to focus their critique on religious moderates. The single constituency that could best enlist with these nonbelievers to effect the political changes they seek is, somehow, the one that they attack with the most vitriol.

In contrast to the New Atheists, the majority of atheists and agnostics are thoughtful and moderate individuals, a storied constituency among the advocates of secularism. One of the goals of this book is to disarticulate secularism from atheism so that secularists and atheists can pursue their legitimate and worthy agendas and work together when their interests overlap (which is often).

The second factor dissuading religious moderates from embracing secularism rests on another incorrect assumption: that secularism is the equivalent of total separation of church and state. A small group of historians and legal scholars, most but not all of whom are conservative Christians, have recently challenged the idea that the Constitution guarantees a "wall of separation." We must take them quite seriously for a variety of reasons, not least of which is that they don't make the slapdash assumption that there is a one-to-one correspondence between nonbelief and secularism. For them secularism is a political concept, not a theological one.

These scholars have argued, sometimes convincingly, that the secular worldview was far less central to America's founding than many have assumed. Separation of church and state, they claim, is a legal fiction – a misreading of the First Amendment initially propagated by the U.S. Supreme Court in the mid-twentieth century. For its own survival, advocates of secularism need to think through this critique quite carefully, as religious moderates (a potential ally with an immense constituency) tend to dislike radical programs of church-state separation.

Secularism will best be served by a coalition composed of two broad constituencies who, for various reasons, are hard to bring together. Nonbelievers are the smaller of the two. This group is filled with intelligent and well-educated individuals of a complex variety often misrepresented by the single-minded extreme atheists. The larger faction comprises believers who, *whether they know it or not,* live by secular or secularish ideals. This group is disorganized, rather listless politically, and by disposition allergic to the sort of sound and fury propagated by the New Atheists. The polemics of the New Atheists have driven a wedge between these potential allies.

The moderates, for their part, are in a most uncomfortable position. On the one side, their Revivalist brethren refer to them as backsliders and whoremongers. On the other, the New Atheists and their epigones ridicule them as clueless dupes in cahoots with the Revivalists. As one scholar complained, secularists push "religious moderates into the arms of their extremist brethren."[28]

This is why the crisis of definition alluded to earlier must be rectified. If no one actually knows what secularism actually is, how can a secular politics gain any traction in this country? If secularism equals Nazism or Stalinism, well, who in their right mind would want to join that club? If secularism is just a form of extreme atheism that snickers at religion's dumb show, then the moderates will not buy in. And if secularism's sole policy prescription is the hermetic walling off of church from state, many moderates will say thanks, but no thanks.

How'd They Do That?

"*I can't believe they did that! How'd they do that? I thought we had separation of church and state in this country?*" Consider this soliloquy to be American secularism's anguished Cry of Dereliction. Again and again, Revivalist activism leaves secularists stupefied and slack-jawed.

How'd Revivalists seize control of the Texas State Board of Education in 2011, laying siege to Thomas Jefferson's wall of separation along the way? How'd they introduce bills on the floor of Congress, attempting to nullify the Establishment Clause? How'd they set up an office within the federal government that forklifts billions of taxpayer dollars into the coffers of "faith-based providers" whose primary course of therapy is prayer and Bible reading? How'd they manage to co-opt major presidential debates, such as the one wherein Pastor Rick Warren grilled the candidates Obama and McCain in 2008? How'd American secularists let any of that happen?

How indeed? Making sense of how secularism was picked apart by the Revival is a major component of the story this book will tell. In order to do that, however, we first need to take a long step back and understand what precisely it was that the Revivalists picked apart.

Which brings us back to the question that haunts our subject matter: what are the core principles of the secular vision in this country? It is difficult to answer because of the dearth of serious big-picture studies devoted to the complex historical development of American secularism. The question of how it emerged, crested, and faded is virtually unexplored. Scholars, for their part, tend to concentrate far more on something called "secularization" than on secularism (and many who do the latter are working within the cancer and corpse paradigms discussed earlier).[29]

In fact, there is no consensus nor master scholarly narrative about the origin of American secularism or how it developed across the history of the republic. Its roots and trajectory remain largely unexamined. This book will trace a very broad outline of those origins in order to initiate a frank, cellulite-and-all discussion about the fundamentals of the secular vision.

We will start by tracing the early modern genealogy of the secular idea. If American secularism was to have a Mount Rushmore enshrining its grandees, where might we situate that peculiar hypothetical monument – Washington Square Park in Manhattan? Whose grizzled countenances would appear on it? We nominate Martin Luther, Roger Williams, John Locke, Thomas Jefferson, and James

Madison for this honor. These five men are the major builders and heroes of the American secular vision that has left its imprint on our politics and culture.

These architects are an odd lot to be sure. Much about them and their relation to secularism is peculiar and paradoxical. Take, for example, Martin Luther (1483–1546). The Father of the Reformation and Protestantism once recounted the following story: "When I awoke last night, the Devil came and wanted to debate with me; he rebuked and reproached me, arguing that I was a sinner. To this I replied: Tell me something new, Devil!"[30] And he, gentle readers, is the father of modern secularism! Despite being fanatically religious, Luther put into play the basic, raw justifications for keeping religion out of the state (and vice versa). Those who equate secularism with atheism are now cordially invited to reassess everything they know about the subject.

The next major architect is Roger Williams (1603–1683). The philosopher Martha Nussbaum, in her excellent book *Liberty of Conscience* – one of the rare scholarly big-picture studies of American secularism of considerable scope – speaks of Williams as a person whose primary concern was "to protect the individual conscience."[31] For Williams that sacred freedom can be protected only by a "civil state that is not religious in character and that does not make laws regarding religion."

The basic secular insight that governments should not get ensconced in religion, and vice versa, was a veritable obsession for this man, one of history's most radical proponents of religious freedom (and by many accounts an insufferable self-righteous bore).[32] According to Williams, Jesus himself professed "that no civil magistrate, no king, nor Caesar, have any power over the souls or consciences of their subjects, in the matters of God."[33]

The cantankerous Williams was a child of Luther's Reformation.[34] While it is fashionable for professors on the Left to tar secularism as an "Enlightenment project" (a term of abuse in these quarters), the truth is, secularism is an odd hybrid. Its roots are sunk in *both* Reformation and Enlightenment soil.

Speaking of the Enlightenment, this brings us to our third architect, John Locke (1632–1704). This great philosopher refined and liberalized the secular intuition of Luther (it is likely that he had also read Williams).[35] From his exile in Holland he penned the resplendent *Letter Concerning Toleration,* an impassioned plea against what we would today call an establishment of religion (a plea, as one scholar points out, that never stopped Locke from supporting an establishment of religion in England).[36] Be that as it may, Locke followed both Williams and Luther in espousing the idea that "man has a natural right to a freedom of conscience."[37] He also followed his predecessors in arguing that church and state would mutually benefit by staying clear of each other.

The towering Locke exerted a formidable influence on Thomas Jefferson (1743–1826) and James Madison (1751–1836)—and both were quick to acknowledge it. Jefferson and the Founders aspired to take the most radical edge of the European Enlightenment and establish it as the *starting point* for the new republic. "But where he stopped short," exclaimed Jefferson in homage to Locke, "we may go on."[38]

The five architects did not agree on every point. But each placed a premium (in their theory but not always in their practice) on the building blocks of that political philosophy about governance that we refer to as secularism. These rudimentary ideas include a warning about the dangers of bringing religion and governmental power into proximity, the celebration of religious freedom, the emphasis on the need for social order to ensure proper communion with the divine, and the idea that all religious groups must be equal in the eyes of the state.

There are other luminaries, of course. The tourist office at the base of the secular Mount Rushmore would surely, in its Premodern Wing, display images of Saint Paul and Saint Augustine. Down the hall in the Victorian Pavilion we would find statues of George Jacob Holyoake, John Stuart Mill, and Charles Bradlaugh. A French Mount Rushmore honoring its variant of secularism, *laïcité,* would honor Voltaire, Jean-Jacques Rousseau, the Enlightenment philosophers, and many others who led us to respect the importance of freedom *from* religion. A rogues' gallery would be set up to examine Lenin, Stalin, and other prelates of Soviet secularism. We will meet them all in due course.

As for American secularism – by studying its very peculiar rise, its short and mild golden age in the mid-twentieth century, and its spectacular fall, we can find clues as to how it might be resurrected. Additional helpful hints will emerge from an investigation of the way this rapid and surprising collapse coincided with the rise of the Revival.

Having established what secularism is not, or is no longer, we need to think of a way forward. This means thinking about how to *be* secular, which is a good deal less complex than many people assume. Secularism is not some esoteric system of radical dogmas beloved of only egghead academics and left-wing elites. Secularism does not require a massive program of Maoist-style reeducation in order to be understood.

Rather, secularism is more or less what most people in liberal democracies – religious and nonreligious – already believe to be true and just. Many groups, be they made up of Christians, Jews, Muslims, or nonbelievers, are already speaking the poetry of secularism. The trick lies in getting the believers to see themselves, in some small but significant way, as secular or secularish. Only when that occurs will it be possible to unite them in action.

How to Be Secular aims to reinvigorate a movement in a state of exhaustion, denial, and utter confusion. Shaken to its foundations by the Revival, secularism needs to take stock of its predicament, to conceptualize itself anew, to commit itself to a season of rebuilding. This necessitates that we reimagine secularism, that we ask of it new, never-before-heard questions. And, if need be, that we blow it up in the hope that it will rise again.

The task, to quote Augustine, is long and arduous,[39] but nonetheless a worthwhile endeavor. Freedom of and from religion are precious values. Peace among the faithful and the faithless is the hallmark of the just society. The secularish virtues of moderation and toleration are civilized graces worthy of being rehabilitated and defended. And the alternatives to secularism now bandied about by American Revivalists need to be strenuously resisted lest the nation fall away from those principles that have rendered it prosperous, tranquil, and strong.

I

WHAT SECULARISM IS AND ISN'T

1

What Is Secularism? (The Basic Package)

> Secularism is the dream of a minority that wishes to shape the majority in its own image, that wishes to impose its will upon history but lacks the power to do so under a democratically organized polity.
>
> – T. N. MADAN, *"Secularism in Its Place"*

THE PEOPLE WANT religion in civic life! The people want their government to support communities of worship nationwide! The people want a more robust role for faith in public schools and other public spaces! The people want to hear "Merry Christmas," not "Happy Holidays" from clerks down at the mall! What the people *don't* want is, well, secularism. Indeed, its critics charge that secularism is antidemocratic, that it runs roughshod over the will of the people.[1]

Might the critics have a point?

A Christian Revivalist in the United States could cite scads of damning statistics in support of this allegation. Sixty-five percent of Americans think that the Founders intended this to be a "Christian Nation."[2] In the aftermath of the September 11 attacks, nearly 70 percent told pollsters that they were praying more.[3] Around 91 percent believed the words "under God" should remain in the Pledge of Allegiance.[4] As for "Merry Christmas" or "Happy Holidays" – 72 percent preferred the less secular salutation.[5]

These gaudy numbers account for why "*Put it to a vote!*" is the new war cry of Revivalists far and wide. Religious traditionalists from South Carolina to Syria have become much more democracy-lovin' of late. They are cognizant of their numerical advantage. A simple referendum, they maintain, would make it abundantly clear that secularism lacks broad appeal, that secularism is apartheid with lipstick.

Overheated as these claims might be, secularists should take them seriously. Secularism is usually about as popular as taxes (an intriguing analogy, which we shall probe anon). With the exception of France, one would be hard-pressed to find a country on earth where a large number of citizens get enthusiastic about secularism.[6]

The majority Sunni Muslim population in Syria, for example, certainly does not like the tottering "quasi-secular" authoritarian regime of the Assad family, who are Alawites (a small breakaway Shiite group considered heretical by the Sunnis).[7] Hindu nationalists in India are none too enthralled with a secular government they feel grants way too many advantages to India's Muslim minority.[8] Nor have Islamists in Egypt warmed to the suggestion that the post-Mubarak state should be secular. "The ballot boxes will decide who will win" – this was the cheery prediction of a recently freed Islamist implicated in the assassination of Anwar el-Sadat.[9] In the view of groups like the Muslim Brotherhood, Egypt would do best if governed by sharia law.[10]

"*Just put it to a vote!*" say the Revivalists around the globe, and then they ask, "*Why is secularism so afraid of the democratic process?*" And with that taunting and tantalizing prompt we are ready to confront some hard, albeit interesting truths about secularism. For while we speak offhandedly about "secular liberal democracies," it is important to grasp the following statement: the ideal secular state is not necessarily a direct democracy and occasionally may be less liberal than some suppose. Secularism is a complex project, with complex relations not only to democracy and liberalism but to religion as well.

But first things first. To understand secularism today, we need to understand the fundamental premises, the basic package of secularism as it developed way back in the Reformation (when the modern world was born, even as it came apart) and the Enlightenment (when philosophers tried to make sense of all of that creation and destruction).

The Yin of Order

Secularism is a political philosophy concerned with the best way to govern complex, religiously pluralistic societies. It aims to strike an extraordinarily delicate balance. On the one hand, it wishes to ensure the existence of a stable social order free of religiously themed strife. On the other, it aspires to guarantee citizens as much religious freedom and freedom to be nonreligious as possible.

Order and freedom: those are the yin and yang of the secular vision. Every secular society has to calibrate a functional equilibrium between the two.[11] Successful versions, like France and the United States, have spent centuries fine-tuning workable accommodations. Other societies fail to achieve that homeostasis and frightful consequences ensue.

In most of those failed cases, the state's interest in preserving order overrides its interest in religious liberty. "Maintaining order" becomes an excuse for maintaining the authoritarian domination of one party. This grim scenario has played out in the former Soviet Union, Kemalist Turkey, Ben Ali's Tunisia, and the failed Baathist regimes whose ghosts haunt the Arab Middle East.

Still, the quest to achieve order is and always has been the raison d'être of the secular project. In the chaotic, combustible world of sixteenth- through eighteenth-century western Europe, where secularism arose, the threat of complete social breakdown due to religious violence loomed large. The attempt to manage this problem left a permanent scar on Western civilization. Today's secular liberals sometimes fail to appreciate the role that order and its alehouse crony, coercive state force, must play in their preferred philosophy of governance.

This emphasis on order appears evident in one of the most important texts in the syllabus of secularism. The author of that text might seem the least likely person to share intellectual DNA with today's secular liberals. But, as we've already seen, secularism is complex. In his 1523 work "On Secular Authority," Martin Luther exclaimed, "We have learnt that there must be secular authority on this earth and how a Christian and salutary use may be made of it."[12] Three centuries later, James Madison, midwifing secular principles of government in America, spoke of Luther's "genius and courage." It was Luther, Madison noted, who called attention to the "due distinction . . . between what is due to Caesar and what is due to God."[13]

Our subject is brimming with ironies and here is a hefty one that may give pause to those who view secularism as a form of atheism: one of the major architects of the secular vision was a person who, by the standards of almost any century, could be considered a religious fanatic. Yet as far as zealous Luther was concerned, secularism was a Christian and God-ordained idea!

The reasons for Luther's embrace of secularism are bound up with his fear of disorder, an evil underwritten by the devil. The father of the Reformation viewed the world as chock full of sinners and the damned. "There are always," he lamented, "many more of the wicked than there are of the just."[14] According to Luther, *even most Christians are corrupt.* "All the world is evil," he frets, and "scarcely one human being in a thousand is a true Christian."[15]

Those distressing odds necessitate the firm grip of a strong prince who can keep the reprobates in check. For "where there is no government," Luther opines, "there can be no peace."[16] But Luther didn't go the next step and insist that true Christians *need other true Christians* to establish and administer that government.

It would be optimal, Luther admits, if that scenario did come to pass. A less interesting or less dynamic thinker might have refused to compromise on that issue. But Luther identified *Christian* justifications for accepting the legitimacy of a prince who was not a true Christian. His sublime contribution was to *outsource* the problem of order to the secular prince. Let *him* deal with the thieves and murderers and barbarians at the gate—all the better to let Christians concentrate on Christ.

To understand how bold a move this is, compare Luther's breakthrough with the opinions of his near contemporary, John Calvin (1509–1564). The latter also pondered the question of the secular authority's relation to Christianity. Yet he came up with an answer diametrically opposed to that of Luther: no outsourcing of any sort is imagined. Calvin concludes, "The end of secular government . . . is to foster and protect the external worship of God, defend pure doctrine, and the good condition of the Church."[17] Calvin's "secular" government was astonishingly similar to a religious government.

But let's get back to Luther. Why shouldn't a true Christian be the person who maintains order and why should true Christians acquiesce to such an individual? Here Luther had some ace New Testament proof texts at his disposal. For starters, Christians are enjoined to "render . . . unto Caesar."[18] They are not interested in worldly power or order-maintenance activities and hence leave them to others. Luther invokes the well-known Romans 13:1–3, wherein Paul exclaims, "Let every person be subject to the governing authorities; for there is no authority except from God, and those authorities that exist have been instituted by God."[19]

Based on this strand of New Testament thought, it is appropriate for a person to display obedience to rulers.[20] A Christian willingly submits to the powers that be. Luther reminds us that the scriptures teach that "even though the powers are evil or unbelieving . . . their order and power are good and of God."[21] It is not for nothing that, centuries later, Jean-Jacques Rousseau would complain that "Christianity preaches only servitude and submission. Its spirit is too favorable to tyranny for tyranny not to take advantage of it."[22]

Another reason for not wanting true Christians to govern is that this act degrades the soul. Rulers, Luther reasons, are entrusted with performing unchristian, albeit necessary, tasks (such as punishing criminals, collecting taxes, waging war). As such, it is better to delegate the job to a secular prince. In the best of all worlds that secular prince would also be a true Christian.[23] Yet Luther is a realist of sorts; he appears resigned to the fact that princes do not typically achieve their position by virtue of their piety and grace.

The prince, then, has a God-ordained job that "has a proper and useful place in the Christian Community."[24] But here is the rub and the starting point of modern secularism: the prince must limit his jurisdiction strictly to taxes, warfare, public order – what Luther termed "outward, earthly matters."[25] As for the "internal" realm of faith, of the soul, of belief – the prince has no business meddling in that sacred domain. That restriction, that fencing off of the heart from what Luther called "the Sword" (meaning the state), is one of the staples of the secular idea.

John Locke's Memories

If Luther is John, then Locke is Paul – the reader can decide whether this is a biblical or a Beatles metaphor. A century and a half after Luther, John Locke would build off his basic ideas while exiled in Holland.[26] Yet he would suffuse those comparatively crude Reformation intuitions with a vitality and sophistication that befits one of the Enlightenment's exquisite minds. Locke took Luther's insights and adapted them to a much more diverse Europe, a Europe teeming with mutually hostile Christian sects.

Luther craved order, but Locke adds an intriguing new twist: *nothing threatens order like religion.* The threat does not come from a particular religion (though for Locke religions like Catholicism were more subversive than others).[27] Rather, *any* religion that commandeers the wheel of the state will wreak havoc upon order. Further, the state that cannot control animosities between religious groups will effectively sow chaos.

One of the reasons that secularism is a bit of a hard sell in today's United States is because we Americans don't have John Locke's memories. We have never seen our country ripped asunder from within by domestic religious strife. True, nineteenth-century America witnessed murderous urban riots that pitted Protestants against Catholics.[28] In the same period there was no small amount of ugliness aimed at Mormons. More recently, we have witnessed similar extreme situations, such as the government's ill-fated assault on the Branch Davidian compound in 1993, which left eighty-seven people dead.[29]

But this is all a walk in Central Park on a Sunday in mid-May compared to the faith-based carnage witnessed in sixteenth- and seventeenth-century Europe. The post-Luther architects of secularism knew of the slaughter committed in the name of religion. They were familiar with the Saint Bartholomew's massacres of 1572, wherein Catholics killed thousands of Huguenots "without regard to sex or age" in Paris and across France.[30] They knew of the Thirty Years War (1618–1648), during which Protestants and Catholics devastated each other and Germany to boot. Though the body count is difficult to ascertain, the figure is plausibly set at eight million.[31] They knew of Puritan and Anglican hostilities that resulted in the deaths of hundreds of thousands in England, Scotland, and Ireland during the civil wars of 1638–1651.[32] They knew – moving to the micro level – of Quakers being imprisoned, beaten, and hanged in the 1650s and 1660s by Puritans in Massachusetts.[33] They knew of the utter insanity of the Salem witchcraft trials.[34]

They knew (depending, of course, on when they lived) about all these things. And all drew a similar lesson: societies tend to rip themselves apart when the power of a religious institution is aligned with that of the government. From Locke forward, the architects of secularism feared that societies with established religions were structurally unstable. By the time of the newly formed American republic, it was taken for granted that establishments of religion "engender strife" and give rise to "a spirit of pride and tyranny."[35] It was Madison who remarked that "superstition, bigotry, and persecution" were the fruits of the "legal establishment of Christianity."[36]

This connection between religion and disorderliness has left its mark on a few centuries' worth of constitutions and legal codes. Every variant of secularism, from the benign to the beastly, displays this obsession with order. The Charter of Rhode Island and Providence Plantations of 1663, issued by Charles II and approved by Roger Williams, proclaims that no person "shall be any wise molested, punished, disquieted, or called in question, for difference in opinion in matters of religion, [that] do not actually disturb the civil peace of our said colony."[37]

A century later, Maryland's Constitution of 1776 warned of the consequences that result when "under colour of religion, any man shall disturb the good order, peace, or safety of the State, or shall infringe the laws of morality."[38] Article 10 of the French Constitution's Declaration of the Rights of Man and the Citizen of 1789 famously proclaims that "no one ought to be disturbed on account of his opinions, even religious, provided their manifestation does not derange the public order established by law."[39]

Even among the beastly secularisms we find this view of religion enshrined in similar documents. For example, a decree of revolutionary Russia in 1918 stipulates that "the free performance of religious rites shall be granted so long as it does not disturb the public order and infringe upon the rights of the citizens of the Soviet Republic."[40]

Much later, in the context of human rights law, the 1966 International Covenant on Civil and Political Rights reflects the same concern. Article 18.3 reads, "Freedom to manifest one's religion or beliefs may be subject only to such limitations as are prescribed by law and are necessary to protect public safety, order, health, or morals or the fundamental rights and freedoms of others."[41]

Never underestimate a secularist's fear of religiously motivated anarchy. Never underestimate a secularist's desire for order.

The Yang of Freedom of Conscience

Revivalists know their history. They are quite aware of secularism's fetish for order. Armed with research on the failed Soviet system, the Revivalists take up their theme. They complain that secularism detests religion. It wishes to do great harm by tyrannically controlling people of faith in the name of social order. Secularism, they argue, is a form of bigotry or even violence aimed at the devout.

Nothing could be further from the truth. Secularism is not in any way opposed to religion.[42] Rather, it does not approve of certain types of relations between religion and government. Such relations, be they establishments of religion, or theocratic rule, or Caesaropapism (wherein the government "treats ecclesiastic affairs simply as a branch of political administration"), always degrade order *and* religious freedom.[43]

Moving from the state level to the psychological level, one might add that secularism is also suspicious of a certain type of religious mindset. One of Locke's most memorable asides is that "every one is

Orthodox to himself."[44] It is customary, Locke understood, for a believer to believe that his way of worshiping the divine is the right way.

By all means, let the believer believe that! But if this believer has the full coercive power of the government behind him, then trouble lies ahead. After all, this self-righteous believer has a counterpart somewhere who also believes that she alone worships the divine the right way. And if this counterpart has a militia at his disposal – well, the rest is history, quite literally. Secularism's job consists of making sure that these two self-righteous believers get to explore their differences only in places like a seminary lecture hall, a journal of ideas, or an online chat room.

Many contemporary critics are quick to overlook the fact that secularism's suspicion of religion is tempered by an earnest effort to protect religion. In one of his typically learned yet accessible essays, Pope Benedict XVI reflects on the genesis of the secular state. It arose, he argues, by "declaring that God is a private question that does not belong to the public sphere or to the democratic formation of the public will."[45] The pontiff's analysis is not incorrect. It does, though, overlook one major motive for the establishment of secular states: the desire to protect religious freedom.

This desire was essential to Luther's view of secularism. The secular prince controls external affairs but may never interfere in *internal* affairs – questions of faith, musings of the soul, thoughts about God, and so forth. Rulers err, thunders Luther, when they have "the temerity to put themselves in God's place, to make themselves masters of consciences and belief."[46] Put differently, a prince is not to dictate matters of faith to his subjects and that's because a prince has no spiritual authority whatsoever.[47]

This was a precious insight, and not one that fell on deaf ears. Roger Williams in 1644 could quote authorities to the effect that "men's consciences ought in no sort to be violated, urged, or constrained."[48] Locke averred that "the Care of Souls does not belong to the Magistrate . . . the Care therefore of every man's Soul belongs unto himself, and is to be left unto himself."[49] In his Virginia Statute for Religious Freedom, Jefferson intones that coercion is "a departure from the plan of the Holy Author of our religion."[50] James Madison concurred, proclaiming that "the religion . . . of every man must be left to the conviction and conscience of every man."[51]

The architects of secularism each wrestled with the question of why a government should never assume the activities of a church. Why should the prince (of Luther), or king (of Roger Williams), or the civil magistrate (of John Locke), or Commonwealth of Virginia (of Thomas Jefferson), or Congress (of James Madison) not become involved in religion? Their response is striking in its uniformity: so that individuals can have the freedom to know God, to revere God as they see fit, and to conceive of God according to the scruples of their conscience.

The soul, all the architects insisted, needs to be protected from the trespasses of the state, or any other entity, for that matter.[52] Along with the need for order, this craving for freedom is the urge that birthed secularism. Anti-secular religious readers, ask yourselves these questions: Is this urge so sinful? Does it indicate a thoroughgoing hatred of religion?

Secularism, contrary to popular belief, takes religious freedom very seriously. It was born of the idea that a particular type of state (that is, one wherein the governing body espouses no particular religion and treats all religions equally) provides benefits for believers in Christ. A magistrate who doesn't interfere in religion would be doing good by God. Such a hands-off leader, Locke enthused, would grant liberty "to Men in reference to their eternal Salvation."[53]

Locke's Escape Clause

Misconceptions about secularism extend not only to its flaws but to its alleged virtues as well. Much is made of the secular virtue of toleration. Yet the role it plays in the basic package is often poorly understood.

What is the single most important Christian virtue? For someone like Saint Augustine, it may have been love of God. For Luther it might be faith in God. But for Locke, casting a dread glance back at more than a century of religiously inspired slaughter, *the ultimate*

Christian virtue is the ability to extend toleration to religious others.[54]

When Locke exclaimed that "Toleration of Christians in their different Professions of Religion" is the "chief Characteristical Mark of the True Church," he redirected the entire energy of Christian thought.[55] The implications of crowning toleration as the king of the virtues are substantial. Locke altered the future of not only secularism, but Christian theology as well.

If toleration is a Christ-like attribute (and orthodoxy a subjective position), then a just polity must set itself the goal of respecting the conscience of every citizen. Subjects cannot be forced to practice religion or worship God in ways that violate their convictions. One sterling way of accomplishing this is to make sure that the state never endorses nor favors a particular religion. As Locke sees it, any state that establishes an orthodoxy will very soon have a plethora of heresies on its hands.

The converse, that a state with no orthodoxy will have no heretics to burn, stands to reason. Toleration grants citizens sovereign psychic space. They can believe anything they want to believe. Faith is a private matter between mortal and creator. The state is not permitted to trespass on this domain. For Locke, a state must live by and enforce the sacred Christian precept of toleration. Call it the state of grace, if you will.

This means that the secular state's sphere of activity is strictly limited. The architects of secularism spoke with unanimity on this point. "By what right," asks Luther, "does secular authority, in its folly, presume to judge a thing as secret, spiritual, hidden as faith? Each must decide at his own peril what he is to believe."[56] "No one," he avers, "can or should lay down commandments for the soul."[57] Roger Williams likewise pronounced himself "against persecution in cause of conscience" and warned against what he termed "Soule rape."[58] Picking up on Luther's emphasis on the inwardness of faith, Locke judges that "Faith only, and inward Sincerity, are the things that procure acceptance with God."[59]

Locke's influence on Thomas Jefferson's worldview is obvious.[60] One of Jefferson's most famous asides can be found in his "Notes on the State of Virginia" of 1781: "It does me no injury for my neighbor to say there are twenty gods, or no god. It neither picks my pocket nor breaks my leg."[61] Jefferson so values the privacy of human consciousness that he respects the right of individuals to hold patently ridiculous thoughts.

Clearly, the architects were adamant about freedom of conscience. But now we come to the darker side of the equation. The secular state must tolerate all forms of belief, *but it need not tolerate all forms of action based on those beliefs.* Scholars refer to this as the belief/acts dichotomy.[62] Put simply, a religious person in a secular state cannot act on her beliefs, no matter how heartfelt they may be, if the resulting actions contravene the laws of the land.

It is not entirely accurate, then, to say that secularism is tolerant. Secularism, yes, is admirably tolerant of thought—far more tolerant than most political philosophies about governance that have come before or after. But it is not necessarily tolerant of all the acts that may result from religious beliefs. An isolated Bible thumper has every right to believe that paying taxes is an offense against God but has no right to act on the impulse. The ultra-orthodox Jew may believe that pelting a female jogger with a tomato on the Sabbath is a proper way of preserving the sanctity of the Day of Rest. And he will be promptly charged with assault and battery for remaining true to his convictions.[63]

The only interesting question in any discussion of toleration is this: *what cannot be tolerated?* It's a bit of a tautology, but secularism maintains that whatever the state cannot tolerate is what cannot be tolerated.

The United States would not tolerate the Mormon practice of polygamy at the end of the nineteenth century, even though the Constitution promises free exercise of religion.[64] In another famous case that elicited a national outcry (and a stare-down between the judicial and legislative branches), the Supreme Court in 1990 would not uphold a Native American's right to ingest peyote in a religious ritual insofar as it contravened the laws of the state of Oregon. Justice Antonin Scalia uncharacteristically invoked the spirit of secularism by using words from the famed *Reynolds* polygamy case of 1879: "To permit this would be to make the professed doctrines of religious belief superior to the laws of the land, and in effect to permit every citizen to become a law unto himself."[65]

Secularism's near-zero tolerance for disorderly religious acts unfolds into another unflattering truth. The secular vision is statist to the core.[66] In a dispute between the state and religion, the state *always* trumps religion. No matter how orthodox a church might be from its members' point of view, the state views this church as equal to all other religious organizations and inferior to the state.[67] Coercion and force will be used if a religious institution somehow loses sight of that. It is very clear from Locke's writings that "Force belongs wholly to the Civil Magistrate."[68] The church must "contain it self within its own Bounds." This means that the church always "loses" insofar as the state monopolizes the prerogative of violence.[69]

So, is secularism antidemocratic, as so many Revivalists allege and as the pope has intimated? A possible answer may be found in a fascinating and overlooked aside from Locke. Deep in the dense prose of *A Letter Concerning Toleration,* the great philosopher opines: "It appears not that God has ever given any such Authority to one Man over another, as to compell any one to his Religion. Nor can any such Power be vested in the Magistrate by the *consent of the people.*"[70]

What Locke is saying is truly remarkable. Governments cannot force religion on subjects *even if the democratic will of the majority has authorized the powers that be to do precisely that.* On certain issues the will of the people is to be ignored. It doesn't matter, for example, if the majority wants mandatory prayer in public school. As for that Revivalist mentioned earlier, with all of his damning statistics about what the people want, the secularist cordially invites him to place himself either at the back of, or under, the bus.

There are two ways to look at this Lockean escape clause. One is to concede that antidemocratic urges abound in secularism. Thus, the Revivalists are correct in tarring secularism as a form of rule akin to apartheid, in which a minority suppresses the will of the majority (that is, those who want the government to espouse a religion).

A more charitable assessment sees secularism as something *prior* to, or something that undergirds, democracy. It provides the preconditions, or operating platform, for representative – as opposed to direct – democracies. There is no democratic "override" of certain secular precepts because that override would signal the *end* of democracy. Secularism says to democracy, "If you permit disorderly religion into the mechanisms of government, if you let religion run amok in public space, you will soon descend into a chaos that will prohibit all of us from getting to God or making it through the day." Thus, secularism emerges as an unalterable, built-in, preexisting feature of democracy. There is no force-quit button.

A parallel to taxation is apt. There must be taxes if a state is to function properly. Yet only remarkably civic-minded individuals feel cheerful when April 15 rolls around. As the activism of the Tea Party demonstrates, a large body of citizens may oppose laws that are necessary for the preservation of the democratic benefits they enjoy. Secularism too performs a vital, albeit highly unpopular, civic function.

The job of secularism is to maintain order. For the citizen to reap the religious benefits of that order, she must make certain concessions. Call it, if you will, the Secular Compact. The state guarantees order and full freedom of conscience. In return the citizen agrees to curtail her religious practices in accord with the laws of the state. Maybe it's a bit problematic that no citizen has actually ever agreed to that deal, or signed over her rights to the secular state. But secularism responds tersely, arrogantly, and on the basis of centuries of empirical evidence that it is simply better this way.

Why Be Secular?

A blow to secularism's ego has been struck. We have pointed to some of the antidemocratic, illiberal, intolerant, and statist characteristics of the secular enterprise. We have also called attention to secularism's "drinking problem" – its proclivity to become so intoxicated on "order" that it can devolve into authoritarianism.

And since we are airing grievances, why not rehearse another oft-heard complaint? The secular vision, it is frequently charged, is a profoundly Protestant vision. Madison, Jefferson, Locke, Williams – all those fellows were Protestants. Martin Luther was the *founder* of

Protestantism. Ergo, it is often alleged that secularism foists all of its Protestant particularities on pluralistic societies.[71]

It thinks of faith in individual, not communal, terms. It wants religion to be private, not public. It disarticulates belief from actions. Neither Catholicism, nor Judaism, nor Islam, nor Hinduism is at peace with all of these convictions. For some critics secularism is akin to an establishment of (Lutheran) religion! Perhaps that charge is an exaggeration. But it is undeniable that American secularism forces Catholics, Jews, Muslims, and especially evangelicals (whose theological affinities now lie more with Calvin) to make significant concessions.

In light of all the preceding drawbacks, one might plausibly ask, Why be secular at all? Despite its imperfections, there are still many very good reasons to support a secular state. Secularism is a fierce and principled defender of religious liberty—perhaps civilization's best defender of it. The complete right to freedom of conscience was a position unambiguously endorsed by each of the architects. When Roger Williams thundered in 1652 that it was "against the testimony of Christ Jesus for the civil state to impose upon the soul of the people a religion, a worship, a ministry," was he not laying the foundation for a truly grand achievement of civilization?[72] Is there any right that Americans of all persuasions hold dearer than freedom from government interference in the realm of the soul?

Later, when we look at secularism's evolution beyond its "basic package," we will see how those core ideas evolved. Secularism's conception of religious liberty is so capacious that by the mid-twentieth century it would even defend freedom *from* religion. Few political ideologies go to the wall, as it were, to secure the freedom of conscience the way secularism does.

Now, it is perfectly true that religious majorities tend to dislike secularism. This implicitly calls attention to one of its great virtues, namely, its great commitment to protecting religious minorities. Secular states offer minorities much-needed security from the majority and even the state itself. Jews in the United States have been impassioned proponents of secularism (and *shhhh!* Catholics have often defended it too, but no one is allowed to say it). The same can be said of countless non-Christian "others" in this country who fear the consequences of living under a Christian establishment.

In other parts of the world the same logic applies. In the aftermath of the Arab Spring of 2011, the late Pope Shenouda III of the besieged Coptic community called for a secular state in Egypt.[73] The Copts had experienced discrimination and violence at the hands of Islamists during the Mubarak era and after. When religious impulses pervade the public sphere, disorder ensues. Few experience the consequences of that disorder more than religious minorities.

Next in the catechism of secularism comes the undeniable record of "secular peace." Secularism emerged amid the smoldering ruins of European societies riven by religious conflict. Its mission was to ensure domestic tranquility. First, it prevented religious groups from fighting one another to seize control of the state. Next, it made certain that the state would never be a source of coercion or violence in the name of any religion. Finally, it guarded against inter-sectarian violence by granting the state a monopoly on the use of force and making it the sole guarantor of order.

And hasn't religious peace been the dividend of secularism in this country? The United States has a fairly deplorable record on race relations, but on religious matters few nations in the world can match its accomplishments. American Revivalists, somewhat ungraciously, tend to forget the progress they made in the twentieth century when secularism was in the ascendant. It was the period of an almost unprecedented proliferation of theological seminaries, journals, and advances in doctrine.[74] It also witnessed a demographic expansion of evangelical Christianity. Secularism, paradoxically, worked so well at creating a space for religion that it may have sparked the Revivalist fire that would like to reduce it to ashes.

But as long as secularism clings to life, let us never deny its other heroisms.[75] All of the Founders put up a spirited defense of the individual's sovereignty over his own mind and spirit. Consider this the "first" privacy, the first of many rights to privacy that would be fought for over the course of Western history. Granting a citizen the right to think anything she wants is the preamble to other privacies. Our homes are considered off-limits to the authorities as long as we act lawfully. Our sexual activities are not subject to scrutiny. We retain the right to confidentiality in matters discussed with our physician. The state cannot trespass there – but the state certainly would do so if the idea that consciousness is inviolable was not so fundamental a feature of Western thought. Secularism may be statist, but it often circumscribes the rights of the state.

Last of all, secularism provides ordinary citizens with essential civil rights. It ensures that, as bearers of a faith or no faith, they are all equal before the law. It guarantees that the state will never interfere in their lawful religious worship. It has the added benefit of creating order – the very order that all the Founders believed was conducive to the contemplation of God.

All well and good, but, as we are about to see, Revivalists in the United States are not buying in. Not at all. Their resistance is based on a boundless resentment for the prevailing secular policy of balancing freedom and order in America. That policy is known as the separation of church and state. If secularism is to revive itself, it will need to carefully ponder that policy's past, present, and future.

2

Were the Founders Secular?

> The secular humanists may argue that we are a secular nation . . . but we are a Christian nation founded on Christian principles.
>
> – DON MCLEROY, *former member of the Texas State Board of Education*

SECULARISTS WHO DESPAIR over the future of the United States have felt few controversies dampen their patriotic spirit quite as much as the recent shenanigans in the great state of Texas.

On March 12, 2010, the Texas State Board of Education voted 10–5 to make radical changes to its social studies and history curriculum.[1] The proposed new course of instruction featured some intriguing ideological tweaks. Textbooks would forthwith describe the country as a "constitutional republic," not a democratic one.[2] High school students would be taught about the contributions made in the 1980s and 1990s by the Heritage Foundation, the Moral Majority, and the National Rifle Association.[3] The Lone Star State's young charges would learn that Moses was an individual who "informed the American founding documents."[4]

Then they got around to Thomas Jefferson, of all people. Previously, the author of the Declaration of Independence had been included in Texas's required unit about "the impact of Enlightenment ideas . . . on political revolutions from 1750 to the present." But now some on the board recommended a stunning change.[5] The man who gave the world "The Virginia Statute for Religious Freedom" was to be purged from the Enlightenment section of the curriculum.[6] Was that expulsion due to the fact that Jefferson is widely understood to be the father of American separationism? Or was it because, as some on the scene speculated, "Jefferson was a deist?"[7]

When the national media was roused to attention, conservatives on the board sought to clarify their position.[8] It was widely assumed in journalistic accounts that Jefferson had been expunged from the *entire* curriculum. Yet the conservatives assured everyone that Jefferson – who promoted "the ideals of limited federal government and states' rights" – was to be removed only from the Enlightenment module but would remain in others. After all, why would they bear a grudge against a first-generation Tea Partier?[9]

As the controversy escalated, conservative Christian members of the board saw fit to share their opinions on all manner of subjects. Cognizant of the fact that schoolbook decisions made in the Texas market will influence many other states in the Union, the board member Cynthia Dunbar declaimed, "The philosophy of the classroom in one generation will be the philosophy of the government in the next."[10] In so doing, Dunbar exemplifies the truism that Revivalists are not only clever but in this for the long haul. They are well funded, well organized, and focused on their goals – descriptors that at the present time certainly do not apply to their secular adversaries.

Some of her colleagues seized the opportunity to voice the mantras of the Christian Right in the presence of the assembled press corps. "To deny the Judeo-Christian values of our founding fathers," enthused Ken Mercer, "is just a lie to our kids."[11] Don McLeroy, the (recently deposed) leader of the conservative bloc and a dentist by training, observed that "Christianity has had a deep impact on our system. The men who wrote the Constitution were Christians who knew the Bible."[12] David Bradley probably summed up the majority's thinking best when he said, "I reject the notion by the left of a constitutional separation of church and state. I have $1,000 for the charity of your choice if you can find it in the Constitution."[13]

The Texas Revivalists' march on the church of church-state orthodoxy was eventually halted – but not before further traumatizing secularists, who have spent the past decade wondering what is happening to their country and weeping their plaints in dulcet tones to brokenhearted interviewers on NPR. The Jefferson episode led them to ask yet again, "*Isn't our democracy founded on the principle of sep-*

aration of church and state? Isn't there supposed to be, like, a wall set up between religion and government?" At the risk of distressing them even more, we must add that scholars are now intensely debating these very questions.

Legal theorists are questioning whether separation is an idea with genuine constitutional warrant. Historians are wondering if Mr. Jefferson's and Mr. Madison's positions truly represent those of the Founding generation. Still others contend that the American government became wedded to the Jeffersonian idea of the separation of church and state much more recently than is often thought.

We now turn to this recent research, though not in order to lend credence to the position of those Revivalists who wanted to remove Jefferson from the Enlightenment module. Their ideas are generally absurd, abounding in a sort of negative intellectual equity. Rather, these issues must be clarified to ensure the future of secularism. Secularists must gain an accurate understanding of their past (and present). They need to engage the legitimate criticisms that are being launched in their direction. If they don't, they will remain perpetually back on their heels, flabbergasted, and out-generaled by their adversaries – as they were throughout the entire fiasco described above.

Is Separation of Church and State in the Constitution?

Secularists nowadays, understandably, make much of the fact that our Constitution is "godless." They cheerfully invoke Alexander Hamilton's response to a cleric who hectored him about the omission of God's name in the nation's founding document. "I declare we forgot it" was his priceless legendary riposte.[14] Hamilton would become significantly less flip and more religious later in life.[15] Though his quip is a pretty high-performing secular talking point.[16]

Still, the beloved "godless Constitution" argument is somewhat imprecise. Our national charter is not entirely godless. The phrase "the Year of our Lord one thousand seven hundred and Eighty seven" does appear in Article VII.[17] Secularists would be better off pointing to the astonishing lack of biblical citations in the document. For the Founders to have refrained (*pace* Dr. Don McLeroy) from invoking the scriptures in which they were so culturally immersed when *absolutely nothing would have stopped them from doing so* is one of the most convincing proofs that they did not have a "Christian nation" in mind.

In any case, Revivalists over the past few years have been developing many shovel-ready talking points of their own. We have encountered some of the popular buzzwords already: "states' rights," "limited federal government," "Judeo-Christian values." But few arguments are heard as often as the one pertaining to the term *separation* and the fact that it is not present in the Constitution. This is like the Everlasting Gobstopper of Conservative Christian spin – an endless source of polemical pleasure.

We learned this lesson not only in the Texas showdown, but also in the 2010 midterm elections. During a televised debate, Christine O'Donnell, running for the Senate in Delaware, rather fecklessly laid an ambush for her opponent. It appeared that she was trying to goad the Democrat Chris Coons into claiming that the Constitution actually contained the phrase "separation of church and state." Coons, a lawyer, did not take the bait.

Their not altogether enlightening back-and-forth went like this:

O'DONNELL: Let me just clarify, you're telling me that the separation of church and state is found in the First Amendment?

COONS: "Government shall make no establishment of religion."

O'DONNELL: That's in the First Amendment.[18]

Opinions differ as to whether this last remark was intoned as a question or as an affirmation. In any case, O'Donnell's exceedingly loony candidacy was interred a few weeks later on election day (though it was probably fated to defeat long before her foray into Con Law 101). Yet the point that O'Donnell's handlers instructed her to make is correct. Those five words are not, in fact, present in the foundational document. The Bill of Rights, as Coons stated, speaks of (dis)establishment, not separation.

Secularists demur, noting that although separation may not be articulated in the letter of the First Amendment, it is present in the spirit of the text. They point, reasonably, to Thomas Jefferson's famous letter of 1802 to the Danbury Baptists. That correspondence, they argue, makes it unequivocally clear what the nation's Founders had in mind.

In the span of a few hundred words, the Danbury letter belts out the classic basic-package themes we encountered previously. "Religion," Jefferson reasons, "is a matter which lies solely between man and his God."[19] Reminding us about the distinction between belief and deed, he observes that "the legislative powers of government reach actions only, and not opinions."[20]

But the short missive is best known for one line in particular. There, Jefferson reintroduced and rejiggered an old metaphor concerning a wall. Scholars debate its origin. Some believe that Jefferson based it on a figure of speech pronounced almost a century and a half earlier by Roger Williams. The latter spoke of "a gap in the hedge or wall of separation between the garden of the church and the wilderness of the world."[21] Jefferson may, however, have had other sources in mind when he reworked that striking phrase in 1802.[22]

Casting a melancholy glance back at the 1780s, he wrote, "I contemplate with sovereign reverence that act of the whole American people which declared that their legislature would 'make no law respecting an establishment of religion, or prohibiting the free exercise thereof,' thus building a wall of separation between church and state."[23]

With those winged words, the Sage of Monticello would lay the groundwork for all manner of legal mayhem. Revivalists fume that the phrase "wall of separation between church and state" has no binding constitutional legitimacy. Rather, it is an aside made by a president in a private letter. Why base domestic policy on an aside, especially one that Jefferson never, ever repeated again? Why endow this "post hoc Jeffersonian statement" with "virtual constitutional status"?[24]

We'll see how that came about in a moment. But the obvious response here is that the Danbury communication helps us understand the Framers' "original intent." The letter permits us to glimpse what James Madison had in mind when he drafted, redacted, and oversaw floor debates in Congress about the religion clauses of the Bill of Rights.[25]

Conservatives should be amenable to this line of evidence. They lament that an "activist" Court has abandoned the Framers' true vision in an effort to secularize America. Doesn't the Danbury correspondence shed light on Madison's (who actually authored the religion clauses) and Jefferson's thinking on this issue? Isn't that proper methodology for an originalist, who seeks, somehow, to live by the letter of the Constitution?[26]

Conservatives might plausibly return fire by noting that the Danbury letter was written thirteen years *after* the drafting of the First Amendment in 1789. In 1789, some note with truculence, Jefferson was futzing around in Paris (where he served as minister to France).[27] What did he know of the deliberations of the Constitutional Convention of 1787 and the First Federal Congress's floor debates on religion? Why take his account as accurate, since he wasn't an eyewitness?[28]

Secularists can parry lines of attack such as this one. But another critique of the wall metaphor is much more difficult to dismiss. In recent decades a growing contingent of critics has issued a challenge to American Secularism's Official Narrative. That narrative paints separation as the unanimous, unambiguous preference of the nation's Founders, who then hastened to make that preference the engine of civil policy. From 1791 to about the time of the Reverend Jerry Falwell, so goes the narrative, we had total separation of church and state in this country. During that time of peace, the wall was sturdy and high and even adorned with a lovely patch of climbing violet wisteria to awe appreciative onlookers on both sides of that majestic divide.

Critics question many aspects of this narrative, especially its emphasis on the unanimity of the Founders. They charge that Jefferson and Madison advocated a conception of church-state relations that was too extreme for their adversaries to accept. But then these researchers go even further. They argue that the vision of the enlightened gentlemen from Virginia was too extreme even for their *allies*.

This last possibility is raised in the scholarship of Philip Hamburger, a law professor at Columbia University who has written an original and influential history of American separationism. In his book *Separation of Church and State,* Hamburger has noted a small but telling detail about the Danbury letter's afterlife. The Danbury Baptists, known for taking down precise minutes, never published or quoted President Jefferson's correspondence wherein he extolled the virtues of the wall.[29] This is an intriguing oversight. Hamburger speculates that this is because the Baptists were dissenters who shared many of Jefferson's views about establishment, *but not his views on the wall.*

"Separation was not what the Baptists wanted," remarks Hamburger. "It was incompatible with their understanding of the pervasive value of Christianity."[30] Quite to the contrary, Baptists "held that all human beings and all legitimate human institutions, including civil government, had Christian obligations."[31] At the risk of imposing our concept of secularism – the word itself, as we shall see in Chapter 4, did not exist until the 1850s – on the Baptists of the 1800s, we might summarize as follows. For them the proper constitutional equation was *secularism = disestablishment,* not the more radical *secularism = separation,* which Madison and Jefferson espoused.

The distinction between separation and disestablishment is significant, though it is glossed over in contemporary usage. Separation, as its name suggests, denotes a firm division between the government and religion. Disestablishment, by contrast, can connote a variety of different things. The First Amendment decrees that "Congress shall make no law respecting an establishment of religion."[32] One reading of this maddeningly ambiguous phrase (see Chapter 9) is that the national government will not endorse nor celebrate nor support *one particular religion.* Yet a disestablishmentarian could theoretically tolerate a state that endorsed, celebrated, and supported religion *in general* without establishing any one religion.

The Baptists, a "beleaguered religious and political minority" that had run afoul of establishments in both the Old World and the New, ranked among Jefferson's greatest supporters.[33] At the same time, they saw Christian moral scruples as playing a central role in guiding the young nation. These brave dissenters thus couldn't fathom casting God completely out of the political life of the republic. Modest governmental support for religion (that is, Protestantism) *in general* presumably did not upset them. This may explain why they lauded Jefferson for battling establishments in Virginia and elsewhere but stayed silent about his stirring paean to separation in the Danbury letter.

Madison and Jefferson: Outliers?

Yet the Baptists weren't the only ones who didn't see eye to eye with the two intellectuals from Virginia. George Washington, in an apparent dig at Jefferson and Madison, would refer to them as possessing "minds of peculiar structure."[34] When the first president pronounced these words in his Farewell Address – which was partly edited and ghostwritten by Alexander Hamilton – he was challenging what he saw as their insistence that "national morality can prevail in exclusion of religious principle."[35]

That the father of our country did not share the opinions of his two distinguished colleagues is noted by a historian who draws this comparison: "Washington did not think that the state must be 'strictly separated' from religion. He agreed that religious worship was a natural right and that the purpose of government was to secure the rights of man, but he did not translate those general principles into Madison's specific limitations on the powers of governments."[36]

Other major Founders stood at even greater distance from the separationist mentality of those minds of peculiar structure. John Adams, our second president, is one of them. When he was called upon to draft the Massachusetts Constitution in 1779, he gave vent to some extremely nonseparationist sentiments: "It is the right as well as the duty of all men in society, publicly, and at stated seasons to worship the SUPREME BEING, the great Creator and Preserver of the Universe."[37] The text then authorizes the legislature to require of citizens to "make suitable provision, at their own expense, for the institution of the public worship of GOD."[38]

Like many early Americans, Adams was less vexed – unvexed, actually – by *state* establishments. This was, after all, the same Adams who in his diaries of 1774 extolled the virtues of a "mild and equitable establishment" of Protestantism in the state of Massachusetts.[39]

John Adams and Washington, along with others such as Samuel Adams, Oliver Ellsworth, and James Wilson, are seen as exponents of an idea known as "civic republicanism," which is distinct from separationism. Advocates of this perspective sought "to imbue the public square with a common religious ethic and ethos – albeit one less denominationally specific and rigorous than the one countenanced by the Puritans."[40] Civic republicanism could be devastatingly unfair to religious minorities and could verge on becoming an establishment itself. But we shall also see that a modified civic republicanism has made something of a comeback in recent years. It has even developed a variant that could fall under the rubric of secularism.

Then there were those who went even further than civic republicanism. They tacked toward a neo-Puritan conception of American government that is the precursor of the contemporary "Christian nation" position. One writer recently enthused that most of the leading political figures of revolutionary-era America were Protestants who wanted "to be guided not by the stark reason of the late Enlightenment [but by] . . . the intercession of Christ and the Holy Spirit in their private and public lives."[41] The resemblance to Calvin's "secularism" is striking.

The Texas State Board of Education, incidentally, has identified many of these civic republican and neo-Puritan figures as staples of the state curriculum. Thus they advocated for the inclusion of Benjamin Rush, John Hancock, John Jay, John Witherspoon, John Peter Muhlenberg, Charles Carroll, and Jonathan Trumbull in course materials headed for the public schools.

These are the Founders whom Christian conservatives most admire. John Jay, for example, made use of the term "Christian nation," arguing that Americans "prefer Christians for their rulers."[42] John Peter Muhlenberg, an evangelical clergyman elected to the first U.S. Congress, actually had nothing to do with the Declaration of Independence or the Constitution but offers proof that a conservative Christian was present in the early workings of our government.[43]

Jefferson's and Madison's convictions, then, were not synonymous with those of the entire political establishment of the nascent United States, nor with those of the people at large. Jefferson's boast, in the Danbury letter, that the religion clauses were an "act of the whole American people" seems far-fetched. Rather, a variety of perspectives on the proper relation between religion and government certainly existed.

On one end of the spectrum we find the Calvinist- and Puritan-inflected "Christian nation" or "Protestant nation" approaches, which are enjoying an encore in Texas today (though it could be argued that they have never actually disappeared). These shade into the less dogmatic civic republican viewpoint, which acknowledges religion in public life as having a valuable social role without affirming a particular doctrinal or denominational stance. Far from driving faith out of public life, these approaches insist that the former enriches the latter.

Finally, at the other end of the spectrum lie the Enlightenment views of Jefferson and Madison. These positions are the nucleus of the secular basic package. They balance freedom and order while positing that it is best for both government and religion to stay clear of each other and for faith to remain a private matter. If separationism is the precise equivalent of secularism (a contestable view), we must conclude that most of America's Founders were not secular.

The scholarship surveyed here does not necessarily indicate – note this – that the separationist stance of Jefferson was *wrong* or somehow detrimental to the nation's well-being. Rather, it shows that it was a minority position in early America. In the words of one scholar, Jefferson was "outside the mainstream" in his thinking on the role of religion in government.[44] As for Madison, he maintained "a pessimism about the social value of religion . . . so extreme that they separate him from all other Founders, Jefferson included."[45] "Their religious ideas and their views on church-state relations," according to a group of historians, "are among the *least* representative of the founders."[46]

These two iconoclastic thinkers advanced the most compelling political vision of their generation as regards the relation between religion and government. At the same time, they were outliers who had charged far ahead of the pack.

The Birth of Separation of Church and State in the Courts

This brings us to a third challenge to the secular narrative: when did the American government actually start enforcing a policy of separation of church and state?

The surprising answer is that separation did not get much serious judicial play until *the middle of the twentieth century.*[47] For the entire nineteenth century "a de facto establishment of Protestantism" was operative on the state and federal levels.[48] This was the religion of the overwhelming majority of Americans. Protestants simply assumed that "the nation's primary moral values were based on the teachings of Protestant religion."[49]

Public pronouncements of the era drive this point home, often with cringe-inducing chauvinism. Today's Christian nation advocates are always eager to quote a particular remark made, in a widely cited opinion, by Justice Joseph Story, who served on the U.S. Supreme Court from 1811 to 1845. It is likely an accurate reflection of early-nineteenth-century sentiments about the relation between church and state. "The Real object of the [First] Amendment," he commented in 1833, "was not to countenance, much less to advance, Mahometanism, or Judaism, or infidelity, by prostrating Christianity; but to exclude all rivalry among Christian sects, and to prevent any national ecclesiastical establishment."[50]

This astonishing statement buttresses the contention that disestablishment, not separation, was the prevailing reading of the religion clauses in nineteenth-century America – a country, incidentally, experiencing the paroxysms of the religious revival known as the Second Great Awakening. The remark also draws attention to a less-than-subtle insistence that non-Christians, as well as Catholics, were second-class citizens in this Protestant country.

This sort of approach was popular throughout the nineteenth century. In *Church of the Holy Trinity v. United States* in 1892, Justice David Josiah Brewer averred that the United States was a Christian nation.[51] A complex figure, Brewer spent the next few decades touring the country, lecturing on the subject and making hair-raising contradictory pronouncements such as this one: "We enforce no religion; but the voice of the nation from its beginning to the present hour is in accord with the religion of Christ."[52]

Other than making occasional triumphalist references to the Christian character of America, the U.S. Supreme Court in the nineteenth century had little involvement with the religion clauses of the First Amendment. Thus, it had few opportunities to expound on the theme of separation. Disputes involving religious rights and liberties rarely came to its attention from 1789 to 1940. One scholar estimates that over the course of this 150-year period, only twenty-three cases concerned these issues.[53]

The Danbury letter, in fact, was discovered by the Supreme Court a full three-quarters of a century after it was composed, in 1802.[54] This came to pass in the crucial case of *Reynolds v. United States* in 1879, which examined the constitutionality of Mormon polygamy in what was then the territory of Utah.[55] In this, the first major test of the Free Exercise Clause, Chief Justice Morrison Waite cited Jefferson's remark.[56] Emerging from the mothballs, Jefferson's metaphor depicting the wall of separation was suddenly treated as "an authoritative interpretation of the First Amendment."[57]

Even with that bit of judicial resurrection, Jefferson's missive waited another *seven decades* to play a major role in American legal thinking. The prelude to the reemergence of the Danbury letter was the "incorporation," or "absorption," of the First Amendment into the Fourteenth Amendment. The upshot was that the federal Bill of Rights would now apply to the states.[58] Thus, the Free Exercise and Establishment Clauses of the First Amendment, once assumed to apply solely to the federal government, quite literally became the law of the land.

The incorporation doctrine is the bane of judicial conservatives, Revivalists, and Tea Party activists. It is not a coincidence that some on the Texas State Board of Education were railing against church-state separation while promoting states' rights. In their view the two are connected. Secular separationism, as far as Revivalists are concerned, was judicial sleight of hand, an act of federal aggression foisted on the American people by liberal Supreme Court justices manipulating incorporation as a tactic to promote their position. A "constitutional republic," the likes of which high schoolers in Texas will be learning about, respects only the exact words of the original document – activist judges be damned!

The separationist catastrophe set in, according to Revivalists, in the *Everson v. Board of Education* case of 1947. In conservative lore this decision triggered "one of the most radical innovations in American Constitutional law since 1787."[59] At this juncture in history Justice Hugo L. Black, in the majority opinion, gave new burnish to the Jeffersonian metaphor and proclaimed that the "wall must be kept high and impregnable. We could not approve the slightest breach."[60] Thus, 145 years after Danbury and seven decades after *Reynolds,* a new equation, *secularism = separation,* was judicially set in place and took off on a midcentury thrill ride.

And what a ride it was! That period witnessed some truly monumental secular judicial victories. Decisions were taken that prohibited the following items in public schools: "release time" for religious instruction (*McCollum v. Board of Education,* 1948), nondenominational prayer (*Engel v. Vitale,* 1962), and Bible reading to start the day (*Abington School District v. Schempp,* 1963).

Let us never forget the *Lemon v. Kurtzman* case (1971), so inscrutable in certain quarters. It famously ruled that government action must have "a secular legislative purpose" and that there may be no "excessive government entanglement" with religion.[61] The pun is irresistible: *Lemon* leaves a bitter taste in the mouths of Revivalists and other conservatives. These are but a few landmark decisions whose cumulative effect was to make separation a respectable and acceptable principle to guide federal and state governments as they considered the relation between their institutions and religion.

Seventy years after *Everson,* conservatives are still enraged by it. They charge that "the Supreme Court has adopted a constitutional doctrine nationalizing the establishment clause which the Framers of the Bill of Rights and the Fourteenth Amendment clearly rejected."[62] Conservatives believe that the Constitution is their friend. Any "correct" reading of it would undo the secularism (that is, separationism) set in motion by what they consider the extraconstitutional rulings of Justice Black and others.

Secularists should know how to respond to these arguments. They might ask, What might be the alternative to "nationalizing" the Establishment Clause? If this clause didn't apply to the states, what type of America would result? As a major constitutional scholar has noted, without the incorporation doctrine "nothing in the U.S. Constitution would prevent a state from conducting an Inquisition, outlawing an unpopular religious sect, [or] establishing a particular church."[63] Indeed, if states were not subject to the Constitution's religion clauses, South Carolina could declare itself a Baptist state. While a few Baptists might fancy that scenario, how pleased would they be if Utah installed a Mormon establishment or if Catholics dreamed up something similar in Maryland?

The equation *secularism* = *separation* was birthed in the mid-twentieth century, as was the generation of lawmakers that succumbed to its allures. The jurists who staffed the Warren and Burger courts drew up this equation. They leaned liberally, so to speak, on the writings of Jefferson and Madison. Too liberally, according to contemporary critics who accuse these justices of focusing exclusively and irresponsibly on those whom we earlier labeled as "outliers."[64]

Many secularists tend to believe that the wall of separation has been rooted in place since the ink dried on the religion clauses in 1791. The truth of the matter is that judicial developments occurring 150 years later resulted in the erection of this formidable barrier. It was not the 1790s, but the period between the 1940s and the 1970s that brought us legal separation as we know it.

Exile to Canada?

In retrospect, it is safe to say that the Revivalists on the Texas State Board of Education overplayed their hand. The subsequent outcry, during a thirty-day period of public comment, led to a major reversal. After initially exiling Mr. Jefferson from the Enlightenment unit of the curriculum in March 2010, the board reinstated him.[65] In May, the *Dallas Morning News* reported that "board members responded to widespread criticism and placed the nation's third president back into the standards."[66]

Still, many amendments guaranteed to enrage secularists were added to the course of study at the final meeting. Texas students will have to "identify major . . . traditions that informed the American founding, including Judeo-Christian (especially biblical law)."[67] Also, they will learn about how the findings of "the House Un-American Activities Committee (HUAC) . . . were confirmed by the Verona Papers."[68] When all was said and done, McLeroy pronounced himself "very pleased with what we've accomplished."[69]

As well he should be. Once again, aggressive and well-organized Revivalists advanced their agenda while leaving secularists on the defensive, even contemplating self-exile to Canada. The conservatives have not merely equated secularism with evil or the Antichrist; such an accusation would be easy to neutralize. Rather, they have made the case that secularism (as equated with separationism) is a legal and historical *aberration* bereft of constitutional warrant. Their strategy has been to attack the legitimacy of secularism as America's political philosophy about governance.

This tactic needs to be countered by a vigorous and sophisticated secular response. It should incorporate pragmatic concessions, intelligent partnering, cunning legal activism, and clear articulation of secular values and how they harmonize with the very best of American values. A stubborn and simplistic response – to simply double-down and insist, yet again, on total separation of church and state – will create more hard times for secularism. We will get to that vigorous secular response later, but first let us see why doubling down is detrimental to the long-term prospects of secularism.

3

Does Secularism Equal Total Separation of Church and State?

It is well to remind ourselves that the completely secular state does not exist.

– D. E. Smith, *"India as a Secular State"*

To believe that the state can remain totally indifferent to religion is constantly contradicted by reality.

– Nicolas Sarkozy, *La république, les religions, l'ésperance*

Secularism equals separation – this is the equation that has carried the day in contemporary American discussions of religion in public life.[1] It has won over both opponents *and* proponents of secularism. The opponents perceive secularists as taking the principle of separation of church and state to absurd and maniacal lengths. They caricature secularists as First Amendment ambulance chasers, extremists who are hell-bent on suing department store salespersons who wish them a merry Christmas.

Some champions of secularism lend credence to that perception. Let's refer to these secularists, with their desire to place religion and government in discrete, airtight containers, as "total separationists." Few embody the ethos of total separation better than Michael Newdow. A physician by vocation (who later trained as a lawyer at the University of Michigan), the estimable Dr. Newdow has advocated on behalf of a variety of ultra-separationist causes.[2] Before the swearing-in ceremonies of both George W. Bush and Barack Obama, he argued that prayer and reference to religion during their inaugurations should be prohibited.[3] He once sued to remove the statement IN GOD WE TRUST from the nation's currency.[4]

But it was Newdow's ingenious quest to have the Pledge of Allegiance deemed unconstitutional that gained him the most notoriety. The secular activist, who is an atheist, was concerned that his grade-school daughter had to recite the daily invocation that contains the words "under God."[5] Upon appeal in 2002, the Ninth Circuit Court in San Francisco concurred with Newdow's argument, holding that this phrase "violates the Establishment Clause."[6]

The ruling unleashed a raging nor'easter of derision. Or, as *Time* magazine put it, "the most spectacular display of religious fervor in recent memory, particularly among elected officials."[7] Senator Robert Byrd, a Democrat, exclaimed, "I hope the Senate will waste no time in throwing this back in the face of this stupid judge. Stupid. That's what he is, stupid."[8] The Senate then voted 99–0 to "reaffirm the language of the pledge"; the House did the same, 416–3.[9]

By 2004, when the case had trundled along to the Supreme Court, Newdow's crusade had become a national anti-secular rallying point. President Bush lambasted the assault on (under) God. Republican and Democratic representatives followed his lead, as did federal judges, the media, and so forth.

When all was said and done the defendant in *Elk Grove Unified School District v. Newdow* was defeated 8–0 (Justice Scalia, who had *already* criticized the Ninth Circuit Court's verdict, had recused himself).[10] Interestingly, the decision did not hinge on the constitutionality of the pledge, but rather on a technicality. It was ruled that Newdow did not have standing to bring the suit because he had divorced his wife and no longer had legal custody over his daughter. Incidentally, "standing," or the right of a citizen to initiate a lawsuit, is emerging as a new and nearly insurmountable obstacle for secular legal activists.

The precise constitutionality of the words "under God," therefore, was left not directly addressed by the Court (though in their observations on the case the justices – getting into the national spirit of things – did not seem the least bit enthusiastic about Newdow's position).[11] Total separationists, impassioned as ever, drew the conclusion that the door had been left ajar for further legal challenges. Which is, and always will be, the wrong conclusion to draw. This is because total separation, in spite of its seductive simplicity and theoretical charms, is unattained and unattainable in a liberal democracy.

None of this means that secularists ought to submit passively to the current Revivalist onslaught witnessed everywhere from the local school board to the statehouse. Quite the contrary – that onslaught must be checked. But it must be checked by *pragmatic* thought and action. This entails clearly understanding what is possible in America and what is not. The clever secular activist starts from that premise, rather than building castles, or walls of separation, in the sky.

To get a sense of separationism's limits we will focus on two countries. The first is our own. It is a nation where separation has a dubious legal rationale, shaky historical grounding, and little in the way of popular support. The second is France, where secularism – or *laïcité*, as it is known there – has a much firmer constitutional foundation, a lengthy history, and mass appeal. The French *like* their secularism. If any people on earth could completely partition government from religion, it must be the cantankerous, contentious, and anticlerical French.[12]

A look at *laïcité* will also put our own system in perspective. If secularisms could be compared to cheeses, then the American version emerges as awfully mild in comparison to its pungent, rank, and aggressive Gallic cousin.

Jefferson and Madison: The "Unessential Points"

Separationism, as an official policy of the American government, has had three sets of powerful and authoritative champions. The first set comprises Thomas Jefferson and James Madison. They were the "minds of peculiar structure" whose heroisms gave us the concepts of "walling off" and constitutional disestablishment. The second would be the Protestant anti-Catholic xenophobes of the late nineteenth century, whom we will meet later. The third is the Warren and Burger courts of the 1950s through the 1970s. It is by no means a coincidence that the first and third sets are intensely disliked by contemporary Christian Revivalists.

Yet we should recall that although the gentlemen from Virginia and the midcentury jurists favored separation, they neither practiced nor believed in *total separation*. Thomas Jefferson and James Madison were not beyond the occasional jaw-dropping church-state trespass. In their early days of public service and even during their mature years as presidents they breached the wall, sometimes ostentatiously and self-consciously.

The most powerful and highly placed separationists of the era, for example, favored penalties for those who worked on Sunday or otherwise disturbed the spiritual tranquility of that day. Jefferson framed legislation punishing Sabbath breakers, as did Madison, who presented his colleague's bill in the Virginia legislature in 1785 and 1786. In part, that bill reads, "If any person on Sunday shall himself be found labouring at his own or any other trade or calling . . . he shall forfeit the sum of ten shillings for every such offence."[13]

Both men had the nonseparationist habit of attending religious services in the House of Representatives.[14] Jefferson let fledgling churches conduct worship in government buildings.[15] The reasons as to why Jefferson and Madison, of all people, behaved in such uncharacteristically pious ways are a subject of debate. Possible explanations range from "genuine devotion to bald political calculations."[16] As for bald calculations, some observe with knowing nods that Thomas Jefferson first started attending church services in the House of Representatives merely forty-eight hours after penning his Danbury letter! "Jefferson balanced his anticlerical words," writes one suspicious scholar, "with acts of personal religiosity."[17]

President Madison, for his part, set aside days of "public humiliation and prayer," especially during the War of 1812.[18] Does this proclamation sound like the words of a total separationist? "I do therefore recommend the third Thursday in August next as a convenient day to be set apart for the devout purposes of rendering the Sovereign of the Universe and the Benefactor of Mankind the public homage due his Holy Attributes."[19] Madison would, however, express reservations about his actions later in life.[20]

Whether in thought or deed, both men appeared to intuit something about total separation: it is not possible. Today's secularists too must recognize that "impurities" in a secular government are inevitable. Every time they see the word *God* on a dollar bill, they need to relax, breathe deeply, and recall the following words of Madison: "It may not be easy, in every possible case, to trace the line of separation between the rights of religion and the Civil authority with such distinctness as to avoid collisions and doubts on *unessential points.*"[21]

It may not be easy indeed! Two centuries of American history attest to the acuity of Mr. Madison's insight. Some wonder if Madison's use of the phrase "line of separation" is a way of walking back his colleagues' reference to an impermeable wall of separation. Madison's "line," argues one historian, "is fluid, adaptable to changing relationships, and unlike Jefferson's 'wall,' can be overstepped."[22] No matter what the case may be, both of these great statesmen – secular activists, please take note – were willing to compromise on the "unessential points."

In the mid-twentieth century, when the U.S. Supreme Court rediscovered the separationist ethos of our two outliers, it was again understood that total separation was an impossibility. Even Supreme Court justices whose rulings were highly sympathetic to the new equation *secularism = separation* questioned the feasibility of total walling off.

Let us at length cite the words of the colorful and long-serving William Douglas, in an opinion he delivered on a crucial case involving the religion clauses. The *Zorach v. Clauson* (1952) decision permitted students to leave school during regular hours and receive instruction on the grounds of religious facilities. Writing for the majority, Douglas opines that

> the First Amendment, however, *does not say that in every and all respects there shall be a separation of Church and State.* Rather, it studiously defines the manner, the specific ways, in which there shall be no concert or union or dependency one on the other. That is the common sense of the matter. Otherwise the state and religion would be aliens to each other – hostile, suspicious, and even unfriendly. Churches could not be required to pay even property taxes. Municipalities would not be permitted to render police or

> fire protection to religious groups. Policemen who helped parishioners into their places of worship would violate the Constitution. Prayers in our legislative halls; the appeals to the Almighty in the messages of the Chief Executive; the proclamations making Thanksgiving Day a holiday; "so help me God" in our courtroom oaths – these and all other references to the Almighty that run through our laws, our public rituals, our ceremonies would be flouting the First Amendment. A fastidious atheist or agnostic could even object to the supplication with which the Court opens each session: "God save the United States and this Honorable Court."[23]

Rest assured that Michael Newdow will get around to that!

In future cases Justice Douglas would show himself to be among the staunchest defenders of the wall. It would be incorrect to brand him as anti-secular. Rather, he was expressing a truth about the limits of separation. In a country with a population, history, and cultural heritage as religious as this one, complete walling off will never be achieved.

Again and again, the Court has drawn attention to the limits of separation of church and state. A similar opinion came forth in the *Lemon v. Kurtzman* case of 1971, which is considered the high-water mark of American secularism. Yet as Chief Justice Burger remarked back then, "Our prior holdings do not call for total separation between church and state; *total separation is not possible in an absolute sense. Some relationship between government and religious organizations is inevitable* . . . separation, far from being a 'wall,' is a blurred, indistinct, and variable barrier."[24] The same year, Justice Thurgood Marshall warned against taking the wall metaphor "too literally."[25] These observations, from jurists in no way hostile to secularism, should give total separationists pause.

The Roquefort of Laïcité

These remarks on the United States raise an important question: are there any countries out there where total separation prevails? Let us immediately qualify that—are there any *livable* countries where total separation prevails? As we shall see, the wall was cast higher than the heavens in the Soviet Union. But that was not the kind of government usually considered to enrich the lives of its citizens.

The most promising candidate for livable total separation is France. We need not exert ourselves to prove the point that few countries on earth could match its quality of life. As for the total separation part, the French appear to have that covered as well. In 1905, after fierce nationwide conflict, the Third Republic explicitly outlined the centrality of the wall to its vision of government.

French separationism was established in the groundbreaking legislation known as the Law of December 9, 1905, Concerning Separation of Church and State.[26] This document is considered one of the cornerstones of modern French democracy. Its passage marked the turning point in a ferocious battle that had lasted longer than a century. The 1905 law makes separation an *explicit* policy of the French government. In the United States, by contrast, separation is an *inference* drawn by secularists.

Laïcité would achieve its first official recognition in the French constitution of 1946, later reiterated in the 1958 constitution of the Fifth Republic. There, France is described as "indivisible, secular, democratic, and social."[27] Once again: by explicitly articulating the nation's standing as *laïque,* the French went beyond the Americans. We have never pronounced our republic to be either secular, as per the 1958 constitution, or separationist, as per the 1905 law.

The French model has a bit of a reputation and is often singled out as the bad boy of world secularisms. *Laïcité* is often equated with the worst excesses and intolerances of the secular worldview. Its detractors tar it as "militant secular fundamentalism."[28] Others refer to it as American secularism's "militant Latin sibling."[29]

The French government, as most everyone knows, actually dictates to its citizens the kind of religion-related clothes and symbols that can be worn in public schools. The famous "Stasi Commission" of 2003 seconded an earlier recommendation that "conspicuous [religious] symbols" ("*les signes ostensibles*"), such as large crosses, headscarves, or Jewish skullcaps, be banned from public schools.[30] The resulting law of 2004 prohibited "the wearing of symbols or clothes through which students display a conspicuous religious appearance."[31]

More recently, in October 2010, the French implemented the "anti-burqa laws." In regard to religious dress they stipulate that "no one, in a public space, can wear clothing intended to hide [*dissimuler*] one's face."[32] Reactions in the United States often mixed perplexity and outrage.[33] "French politicians," charged the editorial page of the *New York Times,* "seem willfully blind to the violation of individual liberties." "The Taliban," it sneered, "would applaud."[34] In our country a Muslim woman is perfectly entitled to wear a headscarf (*foulard islamique*) or full body covering (*niqab*), and French "intolerance" dumbfounds Americans.

The controversies surrounding veiling erupted in 1989 when two schoolgirls were not permitted to wear the *foulard islamique* to class. Since then Americans have looked askance at their old allies, asking how they could so wantonly ignore the principle of religious liberty.[35] We would, however, urge judgmental Americans to understand the French on their own terms. Their secularism is bound up with their unique past and culture and predicated on completely different political assumptions.

Moreover, we ought to recognize that the French are equally puzzled by American secularism. They believe there is *too much* religious liberty in America. Total separationists in the United States must agree. Surely they see great potential in Gallic *laïcité.* But, as we are about to find out, if they think France is some sort of separationist utopia, they're missing something.

Why France Is Not America (and Vice Versa)

American secularism and French *laïcité* can't really understand each other and a historian might do better than a psychologist to explain their failure to communicate.

Our secularism was greatly influenced by Protestant thinkers such as Luther, Williams, and Locke. By temperament and disposition these men were "religion friendly." Three asterisks hover above this claim. The first asterisk is that these architects of secularism were not necessarily friendly to Catholicism. The second is that their friendliness to their own Protestant religion was not so great that they wished to align it with the state. The third is that they all lived in an era during which atheists were nobody's friend; godlessness was the most deplorable status imaginable.[36]

The French situation is very different. "Religion-friendly" is decidedly not the term we would use to describe those late-eighteenth-century Enlightenment thinkers who were the architects of *laïcité*. They positively loathed Catholicism, which was, in most cases, their religion of birth. Most disliked religion in general as well as religious officials (an attitude known as anticlericalism).

And though most of these figures were deists (believers in a Creator who does not intervene in human affairs), a few dabbled seriously in the dark arts of nonbelief. That brew of skepticism, heresy, and godlessness undergirds *laïcité*. If it has a serrated edge when it comes to religion, the critical ferment of the late eighteenth century was its sharpening stone.

Many intellectuals paved the way for France's revolutionary understanding of religion's role in public life. One thinks of Baron Thiry d'Holbach (1723–1789), who in his work *Le bons sens* shocked his contemporaries with claims like these: "religion is an absurdity" and "the idea of God is an absurdity."[37] One scholar refers to d'Holbach, author of the audacious *The System of Nature*, as "probably the first unequivocally professed atheist in the Western Tradition."[38]

Jean-Jacques Rousseau (1712–1778) casually tossed off anticlerical statements such as "anyone who dares to say 'Outside the church there is no salvation,' should be expelled from the state, unless the state is the church and the prince the pontiff."[39] He could also unspool one-liners that certainly weren't appreciated by priests: "All power comes from God, I agree; but so does every disease, and no one forbids us to summon a physician."[40]

As for the deist Voltaire (1694–1778), he is a compendium of secular wisdom. "If there were only one religion in England," he observed, "there would be danger of despotism." "If there were two," he continued, "they would cut each other's throats, but there are thirty, and they live in peace and happiness."[41]

Of course, French Enlightenment critics had not only the Catholic

Church to contend with, but also its partner in crime, the monarchy. This union of cross and crown is known as the ancien régime. As a general rule of history, all hell breaks loose when a secular movement confronts such a duo (consider, for example, Russia's and Mexico's travails in 1917).[42]

Even when the monarchy disappeared, Catholicism remained the avowed enemy of French secularism. The battle of the "Two Frances," one religious and one *laïque,* has been lengthy, dramatic, and quite often horrifically violent. From the French Revolution of 1789, to the emergence of Napoleon a decade after, to the mass hemorrhaging of the Paris Commune of 1870–1871, to the Dreyfus affair, to the civil unrest that accompanied the 1905 laws – the French came to *laïcité* through "blood and tears."[43]

In the United States – talk about mild cheese! – the battle for secularism began in earnest in the 1940s. On our shores, this struggle has never involved a physical confrontation, but rather a *judicial* one. The final days of the Paris Commune, during which "human life has become so cheap, that a man is shot more readily than a dog," has no parallel in the annals of American secular history.[44] In terms of body count, American secularism has been achieved on the cheap. Instead of spilling blood and tears, secular jurists have submitted amicus curiae briefs and very sharply worded op-eds.

There are many other points of contrast. The American Founders made much of religious liberty. Needless to say, in the crazed decade of the 1790s religious liberty was not a central theme in France. The period gives us laws that read like this: "Every gathering of citizens for the exercise of any worship whatsoever is subject to the surveillance of the constituted authorities."[45] Or we encounter decrees that presage that peculiarly French obsession with public space: "No symbol peculiar to a sect can be put in or upon the outside of a public place, in any manner whatsoever."[46]

This would all be unimaginable in the United States, then and now. Our nation's early childhood was relatively safe, secure, and stable, providing a structured environment for the development of a robust concept of religious freedom. In the 1780s and 1790s the American Framers were "using their words" and peacefully composing their foundational documents.[47] The Anglican Church and England – the closest thing we ever had to an American ancien régime – had been vanquished. Its holdover institution, the Anglican Church in Virginia, had been disestablished by the legislative acumen of Jefferson and Madison.

Aside from these two outliers, there were very few enemies of religion among the Founders' political class (the more radical and anticlerical American deist Thomas Paine was living mostly in France at this time). Besides, it would be an exaggeration to call them enemies of religion. The criticisms by Jefferson and Madison were relatively muted, studied, and gentlemanly in comparison to those being shouted on the other side of the Atlantic.

While the Americans debated, deliberated, and for all we know danced out numbers from the musical *1776,* the revolutionaries across the ocean were up to something entirely different. They were writing and discarding numerous constitutions, guillotining priests, massacring opponents, immersing themselves in foreign and civil wars, and eventually cannibalizing their own leaders. Once again, that sort of childhood leaves its scar on a people. "Nations," Rousseau once observed, "are teachable only in their youth."[48]

While the Americans stressed religious liberty, the French placed more of an emphasis on freedom of conscience. This theme is articulated in one of *laïcité*'s (and world civilization's) foundational texts, the Declaration of the Rights of Man and the Citizen of 1789. Article X proclaims, "No one ought to be disturbed on account of his opinions, even religious, provided their manifestation does not derange the public order established by law."[49]

This cryptic sentence points to two central concerns of the French. One we know well by now: an obsession with order.[50] The second is the individual's freedom of conscience.[51] Henri Pena-Ruiz, one of *laïcité*'s great champions in France today, sees liberty of conscience as among the core virtues of the French secular project.[52] Yet when we think of this idea in the French context, caution must be exerted.

French liberty of conscience is often lumped together with American concepts of religious liberty, but the two are not identical. French liberty of conscience, because of the circumstances in which it arose, is tinctured with the idea that a citizen has a liberty *against* religious indoctrination. The French were permitted, and perhaps even encouraged, by their founders to think and say whatever they wanted about religion. And if they started thinking outside the established Catholic box of the dreaded ancien régime, then *tant mieux,* or all the better. This emphasis on the right of the individual to think freely grows organically in the Gallic *laïque* tradition, with its skeptical roots in the radical Enlightenment.

The early Americans, by contrast, were conducting a grand, civil discussion about religious freedom. "How can we configure our government," they seem to ask, "so that each of our citizens can praise God as he sees fit?" In its finest moments, the American model enjoins and empowers its citizens, be they Anglicans, Congregationalists, Quakers, deists, what have you, to commune with God in accord with the scruples of their conscience. The Framers gave us marching orders, so to speak. They encouraged Americans to find their God. The government, by prohibiting itself from interfering with free exercise, recognizes itself as a *danger* to that noble quest.

Yet there is a downside in the Founders' vision that plagues us to this day: an inability to create a space for nonbelievers. Not everyone has a God to find. Religious liberty as granted by the American Founders doesn't necessarily lead to, or even permit, *nonreligious* liberty – the right to be nonreligious or even sacrilegious. Part of the difficulty, incidentally, of being an American atheist is that there is no phrase in the Constitution that sanctifies this identity. Our essential premise of freedom *from* religion does not forcefully enter American secularism's vocabulary until the late nineteenth century (even though tantalizing hints in this direction were offered by eighteenth-century figures such as Thomas Jefferson and the Baptist minister John Leland).[53]

Thus when in 1985 Justice William Rehnquist, in *Wallace v. Jaffree,* denied that Madison viewed the government as needing to remain neutral on the merits of "religion and irreligion," he was correct.[54] The nonreligious have always had to fight for the right to be considered equal in America. This has much to do with the Constitution's inability to take them into consideration. A legal theorist put it aptly: "France is suspicious of the true believer. The United States is suspicious of the non-believer."[55]

From the French perspective the American model allows its citizens *too much* religious liberty.[56] These Americans, the French could conceivably sniff, aren't even properly secular. The Americans guarantee that their Congress will make no law prohibiting the free exercise of religion. This is all well and good, but where is the essential proviso about curtailing free exercise that threatens public safety? Where does the Constitution call attention to that most secular of concerns, order? Yes. Where does it do that? Surely a religious believer in our country does not have *total* free exercise. The authors of the state constitutions, with their consistent emphasis on order, understood that well.[57] But not our federal Constitution, which can't be bothered to devote more than a few dozen words to the question of religion and is certainly disinclined to probe its dark side.

And while we are on the subject, where is the guarantee of the right of conscience in America's founding documents? (Madison, in his initial draft of the Bill of Rights, penned precisely such a provision: "Nor shall the full and equal rights of conscience be in any manner or in any pretext, infringed." It floated out of the text during House debates.)[58] A French person could legitimately ask, What kind of pious secular flea circus are these Americans running?

The Awesome Power of the French State

All of these considerations lead us to speculate that an American total separationist would be right at home in Paris. With all of its skeptical philosophers, with all of its bloodshed in the name of an ideal, with all of this explicit state support for *laïcité,* one would assume that France would manifest a pretty thoroughgoing form of separationism. This must be the right country for Michael Newdow. If there is any candidate for total (livable) separation, one would assume it would be France.

And one would be wrong. The truth is, the French government, probably more so than our own, is deeply entangled in religion. *Laïcité* is a system that brings the state into far closer proximity to religious groups than its American counterpart. With its Establishment and Free Exercise Clauses, the American government tries to keep its religious subjects at arm's length. The French, in contrast, offer their believers overbearing headlocks. To understand why, we need to briefly acquaint ourselves with that country's post-revolutionary period.

The paroxysms of the 1790s were just an appetizer, an *amuse-gueule,* to the more substantive historical course served up a decade later. In many ways Napoleon reversed the policies of the Revolution. His Concordat of 1801 declaims that "the Roman Catholic and Apostolic religion is the religion of the great majority of French citizens."[59] Napoleon stopped just short of making Catholicism the official religion of France.

Later, in 1802, he rescinded his benevolence toward the Catholic Church with the "Organic Articles,"[60] yet nonetheless he surely tamped down many of the Revolution's most anticlerical initiatives; like any efficient and ruthless leader, he put both the church and its enemies in their proper place.[61] As one French expert describes it, Napoleon's tactic demonstrated that now "the Church is in the State" but "the State is not in the Church."[62]

What Napoleon perfected – and what the French never divested themselves of – was the explicit concept of the supremacy of the state over religion (and everyone else, for that matter). Locke and Luther also posited that supremacy, but both were adamant that the state cannot meddle in the affairs of the church nor the worshiper's soul. Napoleon, however, was not encumbered by that restriction.

His ultra-statism was expressed in his treatment of France's major religious communities. He established what is sometimes known as the "Concordat system," a process through which the state granted recognized status to certain faiths. In 1802 this state seal of approval was extended to Lutherans and Calvinists. Judaism was recognized in 1808. The state doled out privileges to these groups and in return monitored them in ways that would be unthinkable in America.[63]

The Concordat system was meant "to protect, but above all to control religion."[64] Reading through Napoleon's "General Provisions for all the Protestant Communions," one is struck by its disparity with the American model. Churches are prohibited from having any relations with a foreign power. They may not change doctrine or dogma without state approval. "The Council of State," we learn, must be "informed of all the undertakings of the ministers of the sect."[65] Of course, these ministers also became salaried officers of the state.[66]

With this we cut to the essence of *laïcité:* its vestigial revolutionary and Napoleonic desire to *control* religion. As the French political scientist Olivier Roy puts it, "*Laïcité* is, above all, an obsession with religion, and it leads to the desire to legislate about religion instead of accepting true separation."[67] For Roy, "there is no true *laïcité* without a strong state."[68] In a recent study an anthropologist interviewed French political officials only to find them insisting, over and over again, that "state control of religion is intended to guarantee public order."[69] The state has a very explicit jurisdiction over religion in France.

Now we can understand why this form of secularism is so bound up with religion. A state that wishes to control and monitor religion cannot build a wall of separation. Remember the groundbreaking 1905 law concerning the separation of church and state? The storied document opens as follows: "The Republic assures the liberty of conscience. She guarantees the free exercise of religions under the sole restrictions stipulated below as regards the interest of public order."[70]

The next article holds that the "Republic will not recognize, salary, or fund any religion."[71] All of which sounds very separationist. But a few sentences later we receive an indication of the limits of separation: "However, the funding concerning chaplains intended to assure the free exercise of religion in public places such as primary schools, colleges, schools, hospices, asylums, and prisons could be added to these budgets."[72]

In one of the first sentences of the law of separation, fascinatingly enough, *the state acknowledged its responsibility to fund religious chaplains*. Even in the most explicit document of its time separating church and state, we find the state entangled with the church! This is followed by a torrent of legislative provisions regarding the salaries of ministers, ecclesiastical property, regulation of worship, and so forth.[73] If we look at France today we see a government deeply enmeshed with religion in surprising and unexpected ways.

In a throwback to Napoleon's era, the French government recognizes certain religions as legitimate but refuses to recognize others that are thought to be "cult-like" (for example, the Jehovah's Witnesses and Scientology).[74] Conversely, it has formal relations with recognized religious groups. There is a bureau of religions and the government even assigns the minister of the interior to serve as *ministre des cultes,* or minister of religions. (Would a similar cabinet position in the United States pass muster?)[75] This minister is charged with, among other things, formally approving the bishops selected by the Catholic Church.[76]

The French state subsidizes private religious schools (though those schools do not force children to receive religious instruction and they teach the national curriculum).[77] It hires military, prison, and hospital chaplains from each of the recognized religions. A recent news item reported that the government was footing the bill for a pilgrimage to Mecca for French Muslim soldiers.[78]

There is no total separation in France. Rather, the government *regulates* some religious groups and refuses to even recognize others. The French model results in entanglements that would be unimaginable in the United States.[79]

The Stunt Double

Total separation of church and state has no future in America. This is because it has no past in America. And besides, it is impossible to achieve, anyway. Why is that? And why is that the case in France as well?

The original pre-national "ethnic casts" of France and the United States were overwhelmingly composed of Catholic and Protestant actors, respectively. The cultures of the nascent republics were immersed in religious worldviews that stretched back over centuries. In ways that are both visible and invisible, religion saturated both societies.

When the secular vision came to these countries – and by that we do not mean separationism, but rather a deep skepticism about how government and religion would interact – faith-based ways of looking at the world were already deeply entrenched. The Sabbath had been observed on Sunday for a good long time. Christmas had fallen on or around December 25 for centuries. Laws encouraging, for example, monogamy were based on ancient religious convictions. The routines of life were informed by religious ideas and habits that had remained in place for millennia.

Secularism never appears in a void. It emerges in societies that are already steeped in religious tradition. As such, nothing short of a complete, violent overthrow of the existing social order will bring about total separation (and even that, we shall see, is bound to fail). Separationism's greatest enemies are history, culture, and force of habit.

The time has come, then, for a paradigm shift in American secularism. The first step: jettison the old equation *secularism = separation* and the policy of total separation that it prescribes. As the French example teaches us, even stridently secular democratic states with a broad consensus to put religion "in its place" are not able to implement that policy.

Instead, separationism – which is, after all, just one way to implement the secular worldview – can play an important but limited role in promoting the secular vision in America. It can be used as a "stunt double." That is to say, its aim should be to gain attention in the public arena by engaging in political activities that are over-the-top and even somewhat dangerous. For example, its practitioners might threaten to impose strict separation when Revivalists get out of hand. Every political movement has its theatrical posers, its grandstanding rabble-rousers who artificially push the dialogue to the extremes. But they are just one part of the show, and certainly not the main event.

Over the past few decades, though, American secularism has allowed that stunt double to become its central protagonist. Too much of what Americans see of secularism is related to a type of knee-jerk assault on inoffensive forms of religion in the public square. Think of the repeated calls for eliminating the tax exemptions granted to religious institutions (secularists, as we shall see, have been barking up that tree for nearly a century and a half) or the obsession with religious icons in public spaces, such as the cross-shaped structure that emerged from the 9/11 wreckage and is slated to appear in a museum at Ground Zero in New York.[80] As James Madison reminded us, we must not waste time on "unessential points."

This does not invalidate the importance of Michael Newdow's case. He may be, technically speaking, correct. No one doubts the sincerity or intelligence of his intervention. But his quest is futile and counterproductive. Total separation is the first of two nonstarters for American secularism. We now turn to the second.

4

Does Secularism Equal Atheism?

> Secularism leaves the mystery of deity to the chartered imagination of man, and does not attempt to close the door of the future, but holds that the desert of another existence belongs only to those who engage in the service of man in this life.
>
> —George Jacob Holyoake, *English Secularism*

American secularism has lost control of its identity and image. That's because the equation *secularism = atheism* is rapidly gaining market share. It is increasingly employed in popular usage, political analysis, and even scholarly discourse. This formula is muscling out an infinitely more accurate understanding of secularism as a political philosophy about how the state should relate to organized religion.[1] If this association prevails, if *secularism* simply becomes a synonym for *atheism,* then secularism in the United States will go out of business.

Which is fine by the Revivalists and which may account for why they perpetuate this confusion. In these circles *secularism* has become another word for *godlessness.* As one journalist perceptively observes, "secular" is a "code in conservative Christian circles for 'atheist' or even 'God-hating' . . . conjur[ing], in a fresh way, all the demons Christian conservatives have been fighting for more than 30 years: liberalism, sexual permissiveness, and moral lassitude."[2]

Not only foes draw this link, but friends as well. The website of the Secular Coalition for America describes the group as "a 501(c)4 advocacy organization whose purpose is to amplify the diverse and growing voice of the nontheistic community in the United States."[3] This community, it points out, is comprised of "atheists, agnostics, humanists, freethinkers, and other nontheistic Americans."[4] An affiliated organization, the Secular Student Alliance, refers to its mission as "to organize and empower nonreligious students around the country."[5]

Why must so-called *secular* organizations be focused exclusively on nonbelievers? After all, just a few decades back, in secularism's mild separationist golden age, all sorts of religious believers could have been categorized as secularists. The term could refer to a Baptist, a Jew, a progressive Catholic, a Unitarian, and so on. Also, there were secular identities that didn't make any reference to a person's religious belief or lack thereof. A secularist might just as likely have been a public school teacher, a journalist, a civil rights activist, a professor, a Hollywood mogul, a civil libertarian, a pornographer, and so forth. From the 1940s to the 1980s all of the aforementioned groups mobilized on behalf of secular causes, the most prominent being separation of church and state.

Aside from being preposterously imprecise, the equation *secularism = atheism* gravely undermines the potential of secularism as a political movement. It leaves people of faith with little incentive to buy in and reduces secularism's personnel to the size of the tiny American atheist movement.[6]

This equation poses a serious public relations problem as well. The atheist movement is not just small, but it is also among the least popular groups in the United States.[7] A survey in 2007 found that respondents viewed nonbelievers more unfavorably than any other cohort they were asked about. This included Muslims, whom the atheists somehow edged out by eighteen percentage points.[8] If atheists are perceived to "own" secularism, its approval ratings will plummet even further.

This is certainly not the fault of atheists, the vast majority of whom are tolerant, self-critical, and moderate in their outlook (that is, secularish). And were a true secular movement to be forged, it should make the eradication of anti-atheist prejudice integral to its platform. The fact remains, however, that the more secularism becomes narrowly equated with atheism, the less it will be able to forge coalitions and pursue its agenda effectively.

Which brings us, then, to the aforementioned impending bankruptcy. The growing popularity of the *secularism* = *atheism* equation has to do with the advent of a group known as the New Atheists. Incensed by the political and cultural might of the Revivalists, this movement crashed into the public square in 2004.[9] The result included enviable book sales, preposterous polemics, and the almost overnight development of a national media platform.[10]

Yet instead of honing their powers of critique on anti-secular Revivalists, the New Atheists advanced a mixed-martial-arts assault on religion *in general*. They gleefully (and catastrophically) set about pitting nonbelievers against *all* believers. They thus included in their onslaught the one constituency in whose hands the future of secularism lies: religious moderates. The New Atheist creed maintains that moderates are just as dangerous and misguided as their extremist co-religionists.

Here is Sam Harris, offering his characteristically subtle take on the question: "Religious moderates are, in large part, responsible for the religious conflicts in our world, because their beliefs provide the context in which scriptural literalism and religious violence can never be adequately opposed."[11] Richard Dawkins, in *The God Delusion,* includes a self-explanatory section titled "How 'Moderation' in Faith Fosters Fanaticism." "Even mild and moderate religion," he avers, "helps to provide the climate of faith in which extremism naturally flourishes."[12] Surely a school of thought that can't distinguish between a member of the Taliban beheading a journalist and a Methodist running a soup kitchen in Cincinnati is not poised to make the sound policy decisions that accrue to the good of secularism.

The precise relation of atheism to secularism needs to be teased out and explained to the general public. This is actually an old dilemma, one that was debated a century and a half ago. There, the possibility was raised that the passions of extreme atheism tend to muck up the agenda of secularism.

Secularism Born Again?

Opinions differ as to when secularism was born. One approach focuses on Christian premodernity and identifies Paul and Augustine as having laid down the initial tracks of the secular vision. The second, which is preferable, points to the high-speed thought corridor that stretched from the Reformation to the Enlightenment. It was there, in early modernity, that the Luther-Locke-Jefferson line carried the secular vision into the sunlight of Reason.

Others, however, identify a completely different starting point: the winter of 1851, when the Englishman George Jacob Holyoake (1817–1906) recalls having coined the word *secularism* (his contribution being the suffixing of that *-ism* onto the word *secular*). He shared that recollection in a book written forty-five years later, so, unless his memory was flawless, perhaps we should not canonize the precise date.[13] Suffice it to say that secularism experienced a third and auspicious birth sometime during the mid-nineteenth century.

For some it is Holyoake, not Jefferson, nor Locke, nor Luther, who is the true father of secularism. And if secularism suffers from a definitional crisis today, let us note in passing that to him must be ascribed some responsibility for that as well. In his works, such as *The Principles of Secularism* of 1871, he somehow managed to define *secularism* in about a dozen different ways.[14]

This complex figure lived a long and tumultuous life in what must have been a very interesting era to think outside the confines of Christianity. The viselike grip of ecclesiastical control was clearly loosening in Victorian England. This context provided a perfect, though not necessarily risk-free, environment for Holyoake and countless other "infidels" to mount a ferocious attack on the status quo.

In his youth, Holyoake was a relentless critic of Christianity. Like so many Victorian dissenters, he spent time in jail for blaspheming.[15] In 1843 he bitterly, but eloquently, recounted the tale of his imprisonment in his essay "A Short and Easy Method with the Saints." There, the twenty-five-year-old protests that "religion is ever found the mother of mental prostration, and the right arm of political oppression."[16]

Holyoake was understandably enraged at having been thrown in a dungeon for an off-the-cuff poke at Christianity he had made while taking questions after a lecture. He complained that this religion forces the concession of "man's noblest right, the right of *expressing* his opinions."[17] "Infidels have never received anything from Christians," he broods, "but calumny, contempt, insult, imprisonment, and death."[18]

As he aged, however, Holyoake's views would mellow and evolve in ways that make him difficult to categorize. In his early life he threw his lot in with atheism, but one scholar sees him drifting to agnosticism in his old age.[19] Another writer refers to him as a "lukewarm" atheist and notes that Holyoake "was sympathetic to religion."[20] He saw secularism as a worldview or system of ethics that moved toward knowledge of God insofar as it assiduously strived to discover the truth.[21] Holyoake, for his part, endorsed an "atheism of reflection," which "listens reverentially for the voice of God, which weighs carefully the teachings of a thoughtful Theism; but refuses to recognize the officious, incoherent babblement of intolerant or presumptuous men."[22]

This calls attention to a very important truth about self-professed atheists in the nineteenth century and most likely today as well: rather than having a fixed lifelong identity as deniers of God's existence, there is a recurrent fluctuation in their thought.[23] Individual atheists change across the course of a lifetime.

Should this be surprising? People change. Theists change. Atheists change. The latter are not godless every minute of their lives. Nor are the former lacking in doubts. Extreme theists and extreme atheists insist on locking people into one fixed identity. But atheist identity is always in flux.[24] How to be secular? In matters metaphysical, keep an open mind or, as we shall see later, "don't get overwrought."

In any case, Holyoake refined and defended his thought about secularism across more than a half-century of published work. His initial comment from 1851, referenced earlier, maintained that "secular" connoted "principles of conduct, apart from spiritual considerations."[25] Gaining attention and growing in stature, the rising star would publish a book a few years later called *Principles of Secularism*.[26] There he offered up a plethora of definitions. It is hard to pinpoint which understanding of his subject he preferred, but the following would seem to represent his view well: "Secularism is the study of promoting human welfare by material means; measuring human welfare by the utilitarian rule, and making the service of others a duty of life. Secularism relates to the present existence of man

. . . [it is] a series of principles intended for the guidance of those who find Theology indefinite, or inadequate, or deem it unreliable."[27]

Holyoake's sympathetic readers were variously confused, elated, or angered by this definition. Let's begin with confusion: a passenger on the Luther-Locke-Jefferson line might be justifiably flummoxed. Where is the reference to order? the state? the church? In truth, political conceptions of secularism were always an afterthought for Holyoake. He did occasionally contemplate the role of government, as when he wrote, "The State should forbid no religion, impose no religion, teach no religion, pay no religion."[28]

Ethicists, however, might be elated by a definition that placed the accent of secularism on moral behavior instead of politics. Holyoake, as we have seen, spoke of principles that were to guide secularists, what he called "a code of duty pertaining to this life."[29] Foremost among them were the following mantra-like propositions: (1) "the improvement of this life by *material* means," (2) "science is the available Providence of man," and (3) "it is good to do good."[30]

Holyoake's definition(s) not only stressed ethics, but ethics geared to the present. "Secularity," he commented, "draws the line of separation between the things of time and the things of eternity. That is Secular which pertains to this world."[31] The emphasis on ethical action in the here and now is constant through all of Holyoake's thinking. Holyoake went so far as to opine, "Giving an account of ourselves in the whole extent of opinion . . . we should use the word 'Secularist' as best indicating that province of human duty which belongs to this life."[32]

Thus far we have encountered two species of definitions of secularism: the political and the ethical. The political was born of the Reformation and the Enlightenment. It stresses the relation between religious institutions and government. Its bearers were our five visionary architects. The ethical definition was engendered by Holyoake and later on we will note its affinities with the thought of Saint Augustine. We would add, parenthetically, that Holyoake's approach has lived a long and healthy life in dictionaries and encyclopedias in the entry titled "Secularism." Curiously, many reference books tend to favor this ethical definition over the older political one that developed in Christian political philosophy.

Among those who were chagrined by Holyoake's approach would be the small but growing band of Victorian infidels for whom he was a hero and leader. True, Holyoake would earn their respect by championing the cause of freedom of expression (a huge issue for an outspoken population that had a knack for getting thrown in the clink).[33] But why didn't his definition reference infidelity or atheism or agnosticism or some such thing?

Holyoake *intentionally* omitted any reference to atheism in his definition of secularism. To understand why is to glimpse a credible alternative to the extreme forms of atheism that are coming to dominate secularism today.

Charles Bradlaugh and the "War Against Religion"

George Jacob Holyoake had a younger, more charismatic, more incendiary, more radical contemporary. A David to his Saul. A Malcolm X to his Martin Luther King Jr.

His name was Charles Bradlaugh (1833–1891) and his biographer describes him as a "proselytizing atheist" who would shout his unbelief "from town halls and market squares."[34] Bradlaugh acquired a well-earned reputation as one of the most angry, uncompromising atheists in the world.[35] He proudly referred to himself as one of the "rough English skirmishers" in "the great Freethought army."[36] The famed orator was the first president of the National Secular Society. Founded in 1866, it was the flagship organization of Victorian infidelry.[37]

Like Holyoake, Bradlaugh had good reason for thinking ill of Christianity. He too spent time in prison for articulating infidel thoughts. Worse yet, Bradlaugh was literally thrown out of the very parliament to which he had been democratically elected. This sad saga, a sort of national scandal, was drawn out over six years, 1880–1886. The controversy, which transfixed England, centered on how a professed atheist could take a religious oath of office.[38]

In 1870 Bradlaugh and Holyoake held a storied two-night debate in front of a boisterous audience of freethinkers. The relationship between the two men was not without its professional and personal animosities – a state of affairs borne out by their often testy back-and-forth.[39] Aside from the fact that the hearty cheers and laughter of the spectators are recorded in the transcript, their discussion is intriguing because it foreshadows many contemporary tensions involving atheism's relation to secularism. Moreover, Bradlaugh's stem-winders remind us that the extreme atheist worldview has not changed one iota in a century and a half.

Let's start with civility – a subject that often comes up in regard to today's hyper-acerbic New Atheists.[40] Holyoake urges the younger, hotheaded Bradlaugh to forgo his taunts and broadsides aimed at believers. The older man encourages graciousness in debates with religious individuals. He reminds his listeners not to humiliate those with whom they disagree, but to see them as "God-sent."[41]

Bradlaugh responds with heat, throttling his adversary about his aversion to skepticism. The rough English skirmisher spent his life debating theists, discrediting the Bible, staring down self-righteous men of God – what kind of secularism would discourage that sort of action?[42] What kind of secularism wouldn't spend its time criticizing theology and religion? "I submit to you," thunders Bradlaugh, "the province and the duty of Secular criticism is to shake the theological teachings of the world."[43]

Holyoake couldn't have disagreed more. Secularism for him had *nothing* to do with theological teaching or anti-theological teaching. Holyoake, consistent with his 1851 position, insisted that secular criticism shared little common ground with atheism. In terms of the movement's goals, he saw the coupling of the two *isms* as "a disadvantage and a disaster."[44]

Again and again in his career he drove that point home. Holyoake opened his *Principles of Secularism* with an epigraph from Harriet Martineau, a noted freethinker. She wrote, "The adoption of the term Secularism is justified by its including a large number of persons who are not Atheists, and uniting them for action which has secularism for its object, and not atheism."[45] The founder of secularism made this point so often and so vehemently that we are forced to theorize that in England (unlike the United States at the time), the equation *secularism = atheism* was commonplace.[46] Ironically, Holyoake struggled all of his life to correct a misperception of a word he himself invented!

Men like Bradlaugh were infuriated by Holyoake's ambivalence. For Holyoake, theism and atheism should be equal in secularism's eyes. Thus he wrote later in life, "Though respecting the right of the Atheist and Theist to their theories of the origin of nature, the secularist regards them as belonging to the debatable ground of speculation."[47] Atheism, therefore, was just like theism in that its subject consisted of "the unknowable and untraceable." As such, atheism can never be made "the basis of a Secular philosophy of life."[48]

Bradlaugh disagreed in the strongest possible terms. In fact, he couldn't make sense of Holyoake. Was he a non-theist? Is that the same thing as an atheist? Flatly rejecting Holyoake's approach, he argued, "The logical consequence of the acceptance of Secularism must be that the man gets to atheism, if he has brains enough to comprehend."[49]

Whereas Holyoake urges indifference to religion, it is clear that Bradlaugh recommends zeal. "Our real Freethought work," fulminates Bradlaugh, "is war against religion."[50] He encourages the listeners to "do battle with the priesthood until their power is destroyed."[51] Charles Bradlaugh is a sort of nineteenth-century mockup of the New Atheist, albeit one who suffered heroically for his beliefs and had a profound knowledge of that which he was protesting. Today's New Atheism, with its aspirations to war against religion, has been cast in his image.

Among extreme nonbelievers today, Holyoake's conciliatory position is either forgotten or discredited. In fact, New Atheists gleefully tar moderate atheists as "accommodationists" and "faitheists."[52] Yet his softer approach does raise a host of crucial questions for the moderate atheist to ponder. Most important, why did Holyoake so assiduously wish to disassociate secularism from atheism?

In Holyoake's opinion, debates with theologians, enlightening as they may be, distracted secularists from pursuing their ethical obligations in this world and to this world. Secularism should be disinterested in arguments for and against the existence of God, which not only distract, but lead to an intellectual dead end.[53] Indeed, the secularist who spends her days negating Christianity will be left not with a new doctrine but with something that might be called "Not Christianity." Holyoake, for his part, was intrigued by the idea of constructing a new ethical foundation, one that is not built on either Christian dogma or the rejection of Christian dogma.

Yet Holyoake's reasons for rejecting the equation were not merely philosophical but tactical as well. Victorian England was a place where atheists were intensely disliked. Holyoake noted that *atheism* had become "a defiant, militant word."[54] The atheist movement – partly because of its advocacy of freedom of expression, birth control, and workers' rights – had acquired a reputation for immorality (Bradlaugh, understandably, wondered why that spoke badly of atheism).[55] Holyoake did not wish to saddle the fledging secular movement with so many negative perceptions.[56]

But Holyoake's central reason for casting off atheism had to with an issue we will look at later: coalition building. The liberal faiths of his day struck Holyoake as natural allies: "The truth is, that there are liberal theists, liberal believers in another life, liberal believers in God, perfectly willing to unite together with the most extreme thinkers, for secular purposes."[57] Since the word *secularism* was coined, it has been hoped that atheists could work together with theists toward secular goals. That hope remains unrealized.

Holyoake's position did not prevail among the Victorian freethinkers for whom Charles Bradlaugh was the flaming star. Then again, secular movements in Britain never amounted to much. Bradlaugh's National Secular Society, for example, chronically lacked for members. As one scholar of the British scene notes, "The secularists were never able to grow into a mass movement attracting widespread support."[58]

This doesn't mean that secularism's association with extreme atheism was the only reason that secularism has been so marginal in Britain. But it does suggest to contemporary nonbelieving secularists that the strategy of working with religious people remains untested.

The Imam Rauf Controversy

George Jacob Holyoake's lifelong quest to disentangle secularism from atheism was based on a variety of concerns. One of his major fears was that the obsessions of nonbelievers tend to distract them from carrying out the ethical agenda of secularism. In particular, they tend to fixate on the existence of God and other complex theological matters. The response of so-called secular groups to the recent "Ground Zero mosque controversy" illustrates precisely that kind of distraction. It also highlights the need for advocates of secularism to conceptualize their issues outside the framework of extreme atheism.

In May 2010 Imam Feisal Abdul Rauf announced plans to build an Islamic cultural center a few blocks north of the site of the World Trade Center.[59] Dubbed the "Ground Zero mosque" by the media (and known as "the Cordoba Initiative" to its supporters), the project quickly polarized the nation. Opponents of the plan saw insensitivity and Islamic triumphalism run amok. Proponents saw discrimination and Islamophobia.

In the imam's telling, the center would be a locus of reconciliation and healing. It would extol the very best virtues of Islamic toleration and serve as a monument to those who perished in the 9/11 attacks. As he observed, "There is a war going on within Islam between a violent, extremist minority and a moderate majority that condemns terrorism. The center for me is a way to amplify our condemnation of that atrocity and to amplify the moderate voices that reject terrorism and seek mutual understanding and respect with all faiths."[60]

On the basis of such words, members of a secular movement ought to be very interested in exploring partnerships with the imam. After all, the right chords are being struck: "moderate majority," "mutual understanding," "reject terrorism." Secularism, as we understand it, loves that sort of thing.

Not all Americans, however, seemed to take the imam at his word. Some protested the project. Others expressed their dismay to puzzled liberal pollsters at the *New York Times*.[61] With midterm elections looming in November 2010, politicians seized on what seemed to be a promising wedge issue.

At roughly the same time that the imam announced his plan, a huge symbolic rupture occurred in the nation's secular community: the Center for Inquiry, a sort of secular think tank, deposed its longtime leader and founder, Paul Kurtz. The octogenarian philosopher is the towering figure of a school of thought known as secular humanism (see Chapter 11). Although the mainstream media paid no attention to it, Kurtz's ouster from the organization he founded in 1991 sent shockwaves through America's small secular humanist and nonbelief communities.

Kurtz was relieved of his duties in what appeared to be some sort of New Atheist palace coup. Citing "concerns about Dr. Kurtz's day-to-day management of the organization," the Center for Inquiry accepted his resignation in May 2010.[62] Kurtz, for his part, claimed that he was "'shoved on an ice floe" by aggressive atheists who had overtaken the CFI.[63] The sort of antagonisms witnessed in the showdowns between Holyoake and Bradlaugh – the moderate bridge builders versus the pugilistic radicals – continue to fester 150 years later (though as a younger man, Kurtz was less moderate).

As spring turned into summer, the regrouped Center for Inquiry, with its New Atheist leanings, saw fit to chime in on the escalating mosque controversy. It concluded that "there should be no legal impediment to the placement of an Islamic community center near Ground Zero, just as there should be no legal impediment to the placement of a church, temple, or synagogue near Ground Zero."[64]

That sounds very wise and evenhanded and just might have won the CFI a whole bunch of new allies in the faith community. The group showed respect for religious liberty and a healthy appreciation for the value of free exercise in the finest traditions of secularism. It's possible that the CFI, aware of the need to partner with others and build effective political coalitions, wanted to demonstrate commitment to finding common cause with religious people of good conscience.

Let's see how they reached out to potential partners:

> CFI's unequivocal support for the legal right of Muslims to place a community center near Ground Zero does not imply that CFI views the new center as an event to be celebrated. To the contrary, CFI is committed to the position that reason and science, not faith, are needed to address and resolve humanity's problems. All religions share a fundamental flaw: they reflect a mistaken understanding of reality. On balance, CFI does not consider houses of worship to be beneficial to humanity, whether they are built at Ground Zero or elsewhere.[65]

Oh well, so much for the coalition! In plainer terms, the CFI's message might be phrased like this: "Well, you do have the right to build it, but you, like every other religious group, are part of the problem and hence not beneficial to humanity. So go ahead and build your stupid mosque. We will seethe in silence and quietly hope (but not pray) for the end of your faith-based way of life."

Once again a lack of nuance has paralyzed a contemporary secular advocacy group. *All* religions are flawed, according to this statement. Houses of worship, this turgid communication laments, really don't do the species any good.

Interestingly, the CFI was so preoccupied with stressing the pointlessness of religion, and playing arpeggios on New Atheist themes, that it failed to raise a legitimate secular objection to the Cordoba Initiative. Following the aforementioned commitment to freedom of religion, a secularist would have unambiguously supported the right of the imam and his group to construct the center. Once the group acquired the proper permits and licenses, and paid the proper fees and taxes, it's all good. A secular state as well as a secular activist would have no concerns with a religiously themed cultural center built by law-abiding citizens.

Yet the Cordoba Initiative did merit concern because it posed a plausible risk to *order.* Not all who opposed the cultural center did so out of blind prejudice and Islamophobia. The imam's disquieting tendency to equivocate about some of his political views and associations raised legitimate concerns.

In retrospect, Imam Rauf could definitely have used a crisis PR team in the summer of 2010. Each week his critics seemed to raise redder flags higher and higher. As the controversy raged, commentators started looking back carefully over the imam's paper trail. Much was made of his remark to *60 Minutes* years back concerning 9/11: "I wouldn't say that the United States deserved what happened, but United States policies were an accessory to the crime that happened . . . we have been an accessory to a lot of innocent lives dying in the world. In fact, in the most direct sense Osama bin Laden is made in the USA."[66]

This is protected speech. Perhaps it is an unfortunate opinion to be held by a person wanting to build a monument to America on the site of a mass murder. But that's his opinion, and secularism sees opinions as inviolable. In a 2005 lecture in Australia, Rauf spoke at length about moderate Islam and sounded quite convincing and sincere. Yet when a questioner asked him about Muslim extremists, the moderate Rauf said, "How do you tell people whose homes have been destroyed, whose lives have been destroyed, that this does not justify your actions of terrorism. It's hard. Yes, it is true that it does not justify the acts of bombing innocent civilians, that does not solve the problem, but after 50 years of, in many cases, oppression, of US support of authoritarian regimes that have violated human rights in the most heinous of ways, how else do people get attention?"[67]

It is absolutely clear that the imam did not condone bombings carried out by extremists. It is also clear that he sees American foreign policy as a complete disaster, something that creates jihadists. Yet might we suggest to the imam half a million better ways for Muslims aggrieved with the United States to "get attention." In fact, the rest of the Muslim and non-Muslim world has been trying to explain to extremist Muslims that no injustice, perceived or real, merits the wholesale slaughter of innocent civilians as a means of attention getting. But again, Imam Rauf is entitled to his opinion even if it obliquely justifies the rage of extremists.

Things got murkier when the topic of Hamas came up. The imam's refusal to describe Hamas as a terrorist organization certainly raised eyebrows.[68] The Cordoba Initiative's website tried to dispel some of the concerns this raised. In so doing, the same maddening ambiguities and equivocations arose. On the FAQ page, a response to the statement "Imam Feisal Abdul Rauf has not condemned Hamas" draws an exasperatingly legalistic distinction between Hamas as a terrorist group and as a political organization.[69]

The imam, we learn, unambiguously condemns terrorism. Not a word of the response, however, discusses the politics of Hamas. Doing so would be problematic because the express political goal of Hamas is annihilating the state of Israel. When a leader of Hamas praised the idea of building a cultural center near Ground Zero, the imam had no comment – at which juncture the crisis PR team might have urged him to publicly declare, "I am not in agreement with this guy," or something of that variety.[70]

In addition to the ambiguity of the imam's political and religious beliefs, distressing questions were raised about the possibility of Iranian and Saudi funding for the center. The imam did not do much to dispel those rumors either.[71] That countries which have recklessly exported extremist versions of Islam across the world might be involved in this project, at this symbolic location, is not only baffling, but deeply troubling.

Should secularists tolerate the Cordoba Initiative? The refusal to disassociate from Iran and Saudi Arabia is a potential deal breaker. It does imply a future course of action detrimental to the interests of the nation. In January 2011, Imam Rauf's partner in the Cordoba Initiative project, the investor Sharif El-Gamal, announced that "the imam and his wife would no longer raise money for or speak on behalf of the project."[72] Its new leadership needs to categorically refuse to take money from states or groups with extremist views. With such assurances rendered, a secularist could, in good conscience, support the cultural center.

Secularism and Extreme Atheism

Holyoake was certainly correct in insisting that secularism does not equal atheism. He was also justified in understanding that moderate atheism and secularism share many common interests. Nonbelievers have an obvious stake in keeping religion off their backs. Secularism provides tremendous resources for helping them achieve that goal since one of its central tenets is freedom *from* religion.

In turn, secularism is enriched by the cultural sophistication that often correlates with being a nonbeliever. Highly educated and well read, atheists can be assets to the cause, valued allies on a par with religious minorities and members of the liberal wing of the larger faith groups. Secularism's problem is with *extreme* atheism. Far from synonymous with the secular, extreme atheism is dangerously anti-secular. Secularism is predicated on full freedom of conscience, a belief that does not jibe well with Bradlaugh's "war" on religion or Sam Harris's wish for the "End of Faith."

Secularism is a philosophy about governance that posits a church and a state; it works with these two basic units of analysis. It cannot tolerate the lack of toleration witnessed among militant atheists any more than it can tolerate the similar characteristics among Revivalists. It is very clear that extreme atheists *would rather that the church not exist,* and this makes their inclusion in the secular camp problematic.

New Atheists tend to make grand rhetorical gestures toward that goal, though little indicates they seriously plan on bringing their ideas to fruition. We now turn to some extreme atheists who did precisely that.

5

How Not to Be Secular

"Heave-ho! Heave-ho!" was out of a distant place and time, a spectral residue of those rapturous revolutionary days when everyone craving for change programmatically, naively—madly, unforgivably—underestimates how mankind mangles its noblest ideas and turns them into tragic farce. Heave-ho! Heave-ho!

—PHILIP ROTH, *I Married a Communist*

SECULARISM CAN'T GET OUT from under one endlessly repeated slander: namely, that it was the official ideology of modernity's most tyrannical regimes. The former Pennsylvania senator Rick Santorum has struck this theme with gusto, if not grammar: "I want to remind people of the societies that have been secular in nature. Starting with the French Revolution, moving on to the fascists, and the Nazis and the communists and the Baathists, all of those purely secularists hated religion, tried to crush religion."[1]

Lest it be thought that only ultraconservative politicians make such insightful observations, here is the almost lyrical verdict of Anthony Fisher, the Catholic bishop of Parramatta, Australia: "'Nazism, Stalinism, Pol-Pottery, mass murder, and broken relationships: all promoted by state-imposed atheism or culture-insinuated secularism."[2] Left-wing professors too crack out a little Pol-Pottery of their own: "Of those killed by states, a large majority—at least two-thirds, according to my rough calculation—was killed by secular states. The true heroes of secularism in the last hundred years have been Adolf Hitler, Joseph Stalin, Mao Tse-tung, and Pol Pot."[3]

Typically, secularists dismiss such provocations as utterly preposterous. This reaction is unhelpful for two reasons. First, the *secularism = murderous atheist regime* meme has become a staple of political, religious, and academic discourse. Believing and nonbelieving secularists had better learn how to neutralize this talking point; they'll be hearing a lot of it from both the Left and the Right.

True, many of these associations *are* preposterous. It would come as a surprise to Jews to learn that the Nazis murdered their ancestors in the name of nonbelief or the separation of church and state. An Aryan-inflected racial anti-Semitism guided the Nazis, not the writings of Jefferson or Holyoake. Nor did the military dictatorships that propped up corrupt Baathist regimes across the Arab world do much more than absurdly fetishize the "order" component of the secular vision.

Yet secularists, especially those of the atheist variety, ought to consider one case very carefully. We refer to that miserable specter, Soviet secularism. Unlike Nazism or Baathism, the Soviet model explicitly invoked many of the grand basic-package ideas of secularism. On paper, anyhow, the Communist Party of the USSR talked the talk, like a tribute band belting out secular anthems with little feeling for, or understanding of, the original.

The Bolsheviks who came to power in 1917 drew great inspiration from the French Revolution of 1789. They were cognizant of the groundbreaking French bill of 1905, discussed previously, that greatly secularized the Third Republic.[4] The Communist Party that emerged from the Bolshevik cocoon did many of the things that secular states do – such as separate church from state and relegate religion to the private sphere. The Party made much of order and paid the requisite homage to freedom of religion. As for freedom from religion, well, they did that in spades.

Aside from lip-synching secular words, the Communists did a whole slew of other things – this time with feeling – that rendered the Soviet experiment in secularism a world-historical human rights catastrophe. And it was an infectious catastrophe at that! The Soviet version of secularism was "covered" by Communist successor states in eastern Europe, China, and elsewhere.

"Have we suppressed reactionary clergy?" asked Joseph Stalin, not at all rhetorically, in 1928. His answer: "Yes, we have suppressed them. The trouble is that they are not yet completely liquidated. Antireligious propaganda is the means that ought to bring to a head the liquidation of the reactionary clergy."[5]

Liquidating clergy? Needless to say, this is not a legitimate aspiration of secularism. Nor is the confiscation of church property. Nor is the attempt to establish atheism as the religion of the state. None of the architects of secularism signed up for that.

Still, the USSR did not lack for secular DNA. For those looking to rethink and renew secularism, the Soviet catastrophe is a cautionary tale of what happens when total separation and atheism and utter madness lock arms. Think of it as an instructional video lasting seven decades and featuring the Dialectical Materialism Players, abounding in misery, malice, and absurdity and titled "How Not to Be Secular."

Abolishing Things and Marxist Payback

There are ideas in the writings of Karl Marx that have not made the world a better place. In his youthful essay "On the Jewish Question," written in 1843, he famously called for "the abolishment of religion."[6] This is one of them.

Some contend that Marx should not be held accountable for the Soviet debacle that ensued decades after his death in 1883. They argue that the father of historical materialism was really a softy, a romantic, and a poet. He lived and worked, it is noted, under extremely trying and humanly reduced circumstances. Besides, he rarely wrote about religion, or communism for that matter; most of his writings were about capitalism. Read the multipronged plan at the end of the 1848 "Manifesto of the Communist Party" – nothing about religion or its liquidation there.

Fair enough. But words, especially words about abolishing things (for example, religion, private property, the reactionary bourgeoisie), often have unforeseen consequences. In this case, these sentiments provided future Bolsheviks – who were not softies, romantics, or poets – with immense ideological motivation and justification for crimes against humanity.

In 1905 V. I. Lenin – among Marx's most passionate and least astute readers – referred to religion as "unutterable vileness . . . vileness of the most dangerous kind."[7] An acquaintance recalled that when Lenin "perceived clearly that there was no God he tore the cross violently from his neck, spat upon it contemptuously, and threw it away."[8] Like so many Bolsheviks, Lenin's orientation toward religion – *all* religion – was one of extreme hatred and hostility. Marxist materialism, he insisted, "is absolutely atheistic and positively hostile to all religion."[9] This extreme anti-theist orientation would poison the Soviet secular experiment from the start.

Three interconnected aspects of Soviet secularism rendered it unprecedented and appallingly at peace with liquidating people of faith: (1) it loathed religion in general, (2) it loathed one religion in particular, and (3) it had a full-blown scientific alternative (or so it thought) that would vanquish and replace religion.

The Bolshevik dislike of religion was, as far as these things go, a pretty heady and organizationally stable brew. Previous atheists, such as the Victorian infidels of the nineteenth century, were trailblazers who had relatively limited resources to use in thinking through their nonbelief and its relation to politics. Besides that, they were almost constantly on the defensive, if not on the run. Whether dodging prison sentences or being dragged out of parliament à la Charles Bradlaugh, pre-Soviet freethinkers had few intellectual resources and even fewer legitimate political options.

The victorious Russian revolutionaries of the early twentieth century, by contrast, had it all. They had at their disposal the entirety of previous French and English religious criticism. In addition, they had Marx's writings and the critical maelstrom it unleashed among Russian intellectuals. Whether it was the anarchist Mikhail Bakunin complaining in 1882 about "the triumphant stupidity of faith" or Frederick Engels exhorting his readers to "make atheism a compulsory dogma," Lenin and his cohort had on hand full-blown anti-theist political philosophies to put into play.[10] More important, from the 1920s forward they had the entire coercive power of the state at ir beck and call.

in's readings of Marx led him to see religion as a product of d inequality. What Lenin termed "the fog of religion" would lift as soon as "a great open struggle for the abolition of economic slavery" was consummated.[11] In theory, the enemy per se was not religion. Faith was a sign or symptom of the enemy's power and malfeasance. Marx himself stated this clearly in a famous passage in "The German Ideology," where he argued that religion was dependent on, was a result of, the "material life process."[12] A Communist wages war on unjust economic structures, not church steeples. The steeples ought to fall on their own *after* their material supports within the economy are toppled.

This raises one of those tragic queries in Russian and world history: if religion was just a reflex, a symptom of a deeper economic problem, why did the Soviets so relentlessly attack the effect along with the cause? Why did they spend a good deal of the twentieth century trying to symbolically and physically annihilate religion if it was just a "superstructure," or outgrowth of a more fundamental problem?

Unriddle the mysteries of dialectical materialism! One explanation is simply that the theory *changed.* One commentator, trying to make sense of how economically minded Marxists became so obsessed by religion, concluded that the shift was "an arbitrary and improper application of vulgar Marxist principles."[13] Another opines that the Bolsheviks opportunistically modified Marx's theory by arguing that "the oppressor classes used religion to keep the oppressed subservient. Therefore religion, as a tool of the bourgeoisie, had to be combated."[14] Bulldozing church steeples could now be seen as a blow to those responsible for unjust economic structures. If only the violence had been confined to steeples; by one estimate 100,000 religious leaders were executed between 1937 and 1941.[15]

Which brings us to the one religion that the Bolsheviks hated beyond all others. Many Russian intellectuals of the early twentieth century felt an intense antipathy toward the Orthodox Church. This undoubtedly helps explain why—theory be damned!—they were willing to pulverize the effect as well as the cause. For generations prior to the Revolution of 1917, members of the opposition had been persecuted by the tsarist regime and its ally, the Orthodox Church.

The situation is quite similar to that of revolutionary France. There, the Catholic Church and the monarchy formed a partnership whose corruption and greed fomented the enmity of numerous sectors of society. An ancien régime, we speculated earlier, often elicits a counterforce that may be equally violent and undemocratic in its actions. Of course, Robespierre and other agents of the Terror in France carried on for only about a decade; the Bolsheviks' run lasted more than seventy years.

Reactionary and obscurantist, the Orthodox Church's fealty to repressive tsarist rulers was never forgotten. According to one observer, "the two forces were closely intertwined" throughout the imperial era (1721–1917).[16] The state "gave the Orthodox Church complete support in many essential ways . . . the Church repaid the state by supporting the tsar and his government."[17]

This support entailed priests informing on subversive parishioners and helping snuff out and suppress opposition groups. Since "the committed Marxist of the Soviet variety sees every competing world view as a personal threat," the Orthodox Church's stranglehold on the minds of would-be Party members (that is, all citizens) needed to be liquidated.[18]

Immediately after the October Revolution of 1917, revenge was exacted. The church was forcibly disarticulated from state power. Its houses of worship were nationalized, its icons and images desecrated, its clergy manipulated and martyred. One historian recounts some tales of violence from the early days (1918–1919): "Metropolitan Vladimir of Kiev was mutilated, castrated, and shot, and his corpse was left naked for the public to desecrate . . . Bishop Germogen of Tobolsk, who had voluntarily accompanied the czar into exile, was strapped alive to a paddlewheel of a steamboat and mangled by the rotating blades . . . Archbishop Vasily was crucified and burned."[19] Ergo, the rigorous, dispassionate, scientific principles of Marxist dialectics yielded to what we might label as payback.

Ironically, the drastic reduction of the power of the once-hegemonic church in the period 1917–1939 allowed other religious groups in Russia to breathe freely and reenergize themselves.[20] Muslims, Jews, Catholics, and Jehovah's Witnesses, among others, were the unwitting beneficiaries of the early revolutionaries' desire to settle old scores.

The Party tried to rectify this state of affairs by eventually doubling down on repression and seeking to abolish all religion. The result was a grim game of faith-based whack-a-mole (with the sickle replacing the mallet) played across most of the twentieth century.

Scientific Atheism

The Communists, ever perspicacious and humane, had a plan for bringing about the end of religion. It consisted of spreading the doctrine of "scientific atheism" throughout the land. One scholar notes that the doctrine "was omnipresent in the daily lives of Communist citizens."[21] "The first 70 years of Soviet power," writes another, "saw a sustained offensive against religion on a scale unprecedented in history."[22]

To grasp the extent of this offensive, consider that no fewer than forty-four museums devoted to this endeavor were in operation in 1930.[23] In its first year of existence, nearly "a quarter of a million people had visited Moscow's Central Anti-religious Museum."[24] Exhibits might include a statue of Galileo in chains, informational dioramas depicting "the disproportionate amount of wealth amassed by the Russian Orthodox church," and a painting of the Virgin Mary that wept tears – phony ones, of course, and visitors were shown how the artificial effect was created.[25] An internal document reveals that one office of the Party delivered 120,679 lectures on atheism in 1954 alone![26]

The Soviets deputized numerous groups, committees, and offices to spread the gospel of scientific atheism. The most widely studied is the League of Militant Atheists. Established in 1925, it drew its membership mostly from members of the Communist Party.[27] The league's goal was "to raise the 'dark masses' to higher levels of consciousness and behavior."[28]

This sort of consciousness raising entailed a whole litany of efforts, at once frightful, futile, comic, and occasionally bizarre. The league published a magazine called *Bezbozhnik* (*Godless*) to disseminate its ideas. It devised strategies for bringing the population over to nonbelief. No one can accuse them of not thinking big: in 1932 they hatched an initiative of creating one million atheist cells to spread their anti-Gospel, perhaps "the most insane plan ever produced by a communist brain."[29]

Across Soviet history, atheist Communists, whether affiliated with the league or the countless other units devoted to atheist missionizing, tried all manner of things to eradicate religion. They cracked open the burial vaults of saints in order to demonstrate that their remains did indeed decompose (in contradiction to widely held religious beliefs).[30] They challenged believers to produce miracles on the spot. According to all witnesses, these staged public showdowns only hardened the faith of those in the audience.[31] As one commentator put it, religion had "powers of attraction against which Communist ideology has little to offer."[32]

In one of its more rarefied intellectual endeavors, Soviets even delved into biblical scholarship. The head of the League of Militant Atheists himself devoted numerous volumes to disproving biblical legends.[33] As any biblical scholar knows, this never works. Few believers read scripture in the flat and literal way that some atheists think they do. Every sentence of the Bible can be interpreted in one way or in a hundred other ways. Or simply ignored. Believers will not be shaken to their foundations by the news that one verse contradicts another – if only because some theologian in their tradition noted the same contradiction a thousand years ago and has already come up with a very good explanation for it.[34]

Many have pointed to the sheer fecklessness of the Soviet atheist propaganda machine. The antireligious cadres they sent out to the people knew little or nothing about the religions they were trying to debunk. Nor were their own doctrines particularly coherent. Scientific atheism "confusingly claimed to be a science while abandoning scientific methods altogether."[35] It was "an atheistic monopoly that never questioned itself or addressed the concerns of its would-be converts."[36]

For all their talk of superior intelligence, enforcers of nonbelief in the Soviet Union were apparently the most bumbling, dimwitted ideologues imaginable.[37] This was, after all, a brain trust that agitated "against Christmas trees in the alleged interest of forest protection."[38] The same group repeatedly tried to replace religious holidays with secular ones. Thus the Feast of the Protection on October 14 became Harvest and Collectivization Day.[39]

All these efforts, naturally, did not achieve their goals. Recent studies of scientific atheism have described it as a disorganized, underperforming mess. Fascinatingly, the Party was perennially bickering within its ranks as to how best to convert believers (or whether it was even worth bothering with them). The history of Soviet atheism was riddled by intense, some might even say crippling, disagreements among Communist antireligious functionaries. On the one hand, there were those who wanted to eliminate religion quickly and by force. On the other were those who preferred more gradual, less drastic measures.[40] Factor in the legendary Soviet penchant for bureaucratic infighting and one may understand why scientific atheism was a colossal failure.

Soviet religious policy was disjointed, inconsistent, and completely illogical. It undulated among street hooliganism, legal persecution, and tactical tolerance (and sometimes all three at once).[41] There were moments of relative calm, as during Stalin's stunning and strategic concordat with the Orthodox Church during World War II. Then there were periods of out-and-out persecution, such as the 1920s and the unlivable Khrushchev period of 1959–1964. In the latter the government prevented children from attending Baptist worship services, demoted or fired religious workers, and added a new provision to the criminal code of the Russian Republic (Article 227) stipulating imprisonment for religious activity that was "harmful to health."[42]

In retrospect, the massive propaganda machine of scientific atheism did not work. A survey taken in 1936 revealed that a walloping 56 percent of the population still believed in God![43] And this figure is likely low; after all, who is foolish enough to tell a Stalinist pollster that he reveres the Almighty? The results of this study surfaced only in the 1990s.[44]

Around that time arose another confirmation of the failure of scientific atheism. For when Soviet Communism fell, priests, monks, rabbis, and mystics of all flavors ascended, as if on Jacob's ladders, from every catacomb and beet cellar in eastern Europe and central

Asia. Bloodied, with gauze wrapped around their pummeled heads, they nevertheless returned invigorated, their spirits steeled by the decades they spent in the Kremlin's iron furnace.

The Secular Motions

The Soviets always attempted to present themselves as adhering to the precepts of the noble secular vision. Reading through their legislation on religion, one can often detect shout-outs to long-standing staples of the secular worldview. But the government completely disregarded them in practice, all the while implementing what we could only call "innovations" in church-state relations.

Still, we should note that the USSR never went, let's say, the Albanian route. The Communist regime there actually declared itself an atheist state in 1967 by annulling previous statutes that had granted some small, symbolic modicum of religious liberty in 1949.[45] Those statutes were roughly similar to the extremely ungenerous Soviet laws discussed in this chapter. Yet the USSR never officially pivoted away from defining itself as a secular state to take on the mantle of an atheist state. Indeed, throughout its history it maintained the veneer of secularism.

Like most secularisms, the Soviet model began by moving religion out of the public sphere. The first major legislation on this matter, The Decree on the Separation of Church and State, was delivered on January 23, 1918. Of its thirteen provisions these are the ones of greatest interest to us:

1. The Church is separate from the State.
3. Each citizen may confess any religion or no religion at all . . .
5. The free performance of religious rites shall be granted so long as it does not disturb the public order and infringe upon the rights of the citizens of the Soviet Republic . . .
9. The school shall be separate from the Church. The teaching of religion is prohibited in all state, municipal, or private educational institutions where a general education is given . . .

13. All property belonging to churches and religious associations existing in Russia shall become public property.[46]

As often occurs in documents related to Soviet governance, the laws sound sort of tolerant (as in provisions 3 and 5) until one hits something quite sinister (as in provisions 9 and 13). But even the sort-of-tolerant-sounding tenets mask a troubling reality. For the truth is that the claim of devotion to separation of church and state and to religious freedom was completely disingenuous.

States relegate religion to the private sphere for a variety of reasons. In some cases, this is done to check the power of a majority religion. India, for example, adopted secularism in part to protect Muslims and other minorities from the numerically immense Hindu majority. In multireligious societies where the size of the differing groups is more or less equal, a different justification prevails. Separation is essential to keep the competing actors from gouging one another's eyes out (the case of Communist Yugoslavia comes to mind).[47] The state adopts no single religion so as to remain neutral and to prevent faith-based groups from trying to use government to their advantage.

In the USSR the rationale for disestablishment was somewhat different.

The Communists moved religion out of the public sphere in order to better uproot it for good, to make sure it would *disappear*.[48] Separation, then, was not an end geared toward maximizing order and freedom, as it would be in an authentic secular state. Rather, it was a transitional stage. Marxists love transitional stages! The socialist state, for example, was supposed to "wither away" naturally en route to the final stage of communism. Of course, when it came to religion, the authorities did all in their power to *induce* the withering process. So what we have in the Soviet model is a state committed not to the privatization of religion, but to its obliteration.

When the 1918 decree was enacted, the Bolsheviks were locked in a battle with tsarist Russia and the Orthodox Church. In one sweep the separation of church and state and the secularization of educational institutions (and the eventual confiscation of church property and murder of clergy) were set in motion to pulverize the ancien régime. Later the same decree would help uproot other faiths. Separation of church and state was, in the words of one commentator, a "weapon used by the state."[49]

Separation is again mandated in the infamous Article 124 of Joseph Stalin's constitution of 1936. "In order to ensure to citizens freedom of conscience," reads this document, "the church in the USSR is separated from the state, and the school from the church."[50] This sort-of-tolerant-sounding paragraph then takes a peculiar turn: "Freedom of religious worship and freedom of antireligious propaganda are recognized for all citizens."[51]

Freedom of antireligious propaganda? That's new. In the United States such activities might fall under the rubric of freedom of speech. In the USSR, by contrast, criticizing religion was an explicitly articulated civil right. Were such a right to be granted solely to individual citizens – the Slavophone equivalent of our New Atheists – perhaps we could see this as an odd but harmless provision. The problem is that both the Communist Party and the Soviet government (which were more or less the same thing) were committed to antireligious propaganda. Herein lies the irony, or absurdity, of Soviet-style disestablishment. As one observer has succinctly noted, "The constitution guarantees something [that is, freedom of religion] which it is an important domestic political aim of the state to destroy."[52]

The official Soviet version of secularism claimed to respect religion. But for most of its history it did everything within its power to annihilate it. It pursued its objective with vigor. One analyst sums up the situation in the 1930s under Stalin:

> More than 95 percent of the 100 thousand places of worship that had operated in 1917 were forcibly closed; the ranks of clergymen, numbering approximately 300 thousand before 1917, were decimated by murder, exile, and prison; all religious administrative centers were closed; and all religious publications ceased. By 1939 the Soviet landscape was virtually free of any signs of its great religious past, except for sturdy edifices that were mostly converted to secular uses such as museums and movie houses.[53]

Five Lessons Learned

The Soviet Union: secularism's death star. It is to the architects' vision what the suicide bomber is to the moderate Muslim. It left behind millions of dead, ruined countless cultures, and brought about the almost irreversible decline of societies situated behind the Iron Curtain (and beyond). As for the doctrine of scientific atheism, it now retains less intellectual heft than the teeny-tiniest babushka doll.

Teachable moments of the bleakest variety abound in this story. Secularists, particularly the nonbelieving ones, must extract from it some valuable lessons. The single most important one: *do not fetishize separation of church and state.* Separation *might* be a necessary condition for a healthy society, but it is absolutely not a sufficient condition. A commitment to separation must be soldered to an equally robust commitment to religious liberty. If not, secularism is a pointless and potentially frightening endeavor. The case of the Soviet Union confirms this observation: it was totally committed to separation, but from a human rights perspective (and many other perspectives) it was a total disaster.[54]

Maybe – let's be frank – separationism isn't even a necessary condition for domestic well-being. Nonsecular Anglican Britain, after all, was an infinitely better place to live than secular Russia precisely because of its ability to respect religious freedom. Finland has not only one established church, but two (Lutheran and Orthodox).[55] It is one of the most high-performing welfare states in the world. Scandinavian countries such as Denmark and Norway (Lutheran again) – those darlings of American liberals – both have established churches.[56] And who, other than an anti-tax zealot or a sun worshiper, wouldn't want to live there?

Which brings us to lesson number two: *a secular state cannot espouse a religion.* It seems safe to say that the Communist Party advocated and promulgated the quasi religion of scientific atheism. And it seems safe to say that this was a horrendous project.

Lesson three: *hatred of religion, like hatred of atheism, is an impulse that should be tempered.* The Soviets could not envision a polity where they might live alongside people of faith because such people signified the existence of something they saw as thoroughly, almost diabolically, wrong with the economic order of things. This revulsion for religion – the idea that it is has no integrity, that it is tantamount to, in Christopher Hitchens's term, *poison* – is where Soviet secularism and New Atheism converge. If atheists cannot make peace with the idea of the existence of religion, they will never be able to function in democratic polities or a true secular movement. The same holds true for Revivalists who loathe atheism.

Which brings us to recommend another moral of this story: *nonbelief cannot be spread by force.* The human soul is such a complex mechanism; its carapace resists coercion and tyranny. Soviet atheists tried everything to separate worshipers from their gods. Nothing worked. Actually, the Soviet case raises a more complex possibility: *nonbelief can rarely be spread by persuasion.*

Few empirical studies address the subject, but attempts to "rationally" distance a person from faith do not often work. George Jacob Holyoake, for one, understood that if conversion (to atheism) does occur, it occurs "by slow degrees."[57]

In 1989, toward the end of her long and combative life, America's most infamous atheist, Madalyn Murray O'Hair, set up a booth at the Moscow International Book Fair to sell some of her literature. According to her biographer, the infamous atheist was disheartened by what transpired in front of her eyes. The relentless critic of religion realized that while she was ignored, "the crowd consumed religious literature, including 10,000 free New Testaments."[58]

Extreme atheists can draw their own lessons from this anecdote. One hopes they will achieve insight. Secularists, for their part, must never forget the dangers that result when they align themselves with anti-theist zealots.

II

THE VERY PECULIAR "RISE" AND FALL OF AMERICAN SECULARISM

6

The "Rise" of American Secularism and the Secularish

> Leave the matter of religion to the family altar, the Church, and the private school, supported entirely by private contributions. Keep the Church and State forever separate.
>
> – PRESIDENT ULYSSES GRANT, *September 1875*

ANY CONTEMPORARY OBSERVER of movement secularism in America (which nowadays is virtually identical with movement atheism in America) can't help but notice that the movement is preoccupied with the word-and-thought-defying size of the movement. Secular and atheist organizations are *constantly* referring to their swelling numbers, dynamic growth, unbounded energy, imminent expansion, and so forth.[1]

On the basis of these swaggering self-reports, one might assume that American secularism is a ready-to-rumble juggernaut, its legions poised to put Revivalists in their place. This tendency toward hyperbole was evident in Professor Daniel Dennett's New Atheist manifesto, "The Bright Stuff," which appeared on the op-ed pages of the *New York Times* in 2003.[2] There, the respected cognitive scientist alleged that in the United States there were 27 million "Brights," those who reject supernatural explanations for any sort of natural phenomenon. These Brights were described as a "silent majority." If further stoked by the policies and rhetoric of the Bush administration, they were poised to become "a powerful voice in American political life."[3]

But when election day arrived in 2004, a different powerful voice in American political life made its will known with deafening certainty. These were the "values voters" (that is, white conservative evangelicals with ample quantities of traditionalist Roman Catholics and Mormons mixed in) whose ballot endowed George W. Bush with a second term.[4] When all was said and done, the only question remaining about the so-called Brights was whether Dennett's figure was off by 25 million, 26 million, or more.

Dennett was not alone in making overly optimistic interpretations of demographic data. Susan Jacoby, in her 2004 book *Freethinkers: A History of American Secularism,* alleged that the "secularist minority is much larger than any non-Christian religious group" in the United States.[5] Jacoby was leaning on findings from the 2001 *American Religious Identification Survey.* This research reported that "unchurched" respondents had almost doubled in size, from 8.2 to 14.1 percent, between 1990 and 2001.[6] Thus by 2001 it was found that some 29.5 million people fit the survey's category of "no religion groups."[7] Jacoby drew the conclusion that the burgeoning unchurched cohort constituted would-be secular sisters and brothers in arms.[8] This conclusion is also unwarranted.

Let's look at it this way: A Catholic who tells a pollster that she is a Catholic *is a Catholic.* She is a daughter of an age-old organization with stable beliefs and dogma and mechanisms for promulgating them. She has places to pray, to consult, to dialogue, to query. Catholic places. Places where people recognize one another as Catholics. They do so because the Catholic Church has spent the better part of the past two millennia fashioning people just like them into Catholics with a sense of Catholic identity.

By contrast, a person who tells a pollster that she has "no religion" is, sociologically speaking, *nothing like a Catholic.* She does not belong to an age-old organization with a well-formed creed. She has no central institution where she can consult, or dialogue, or query about her lack of religion. She does not necessarily share anything else in common with all of the other folks who say they have "no religion."

And that's because "no religion" is not a coherent identity with any basis in reality beyond a demographer's clipboard.

She is simply a member of a theoretical aggregate who *might* one day be dragooned into an (atheist) secular movement – were such a movement to have the infrastructure and imagination to entice her. It is a leap of faith to presume that a person who says she has no religion is a secularist in the self-conscious and intense way that Dennett and Jacoby construe their secularist identity. A negative identity is not the same as a positive one.

Movement secularists and atheists wildly overstate their numbers.[9] Yet they must surely recognize that their political clout is limited, if not nonexistent. A statistic such as the recent finding that *not one* of the 535 members of Congress claimed to be an atheist is astonishing.[10] The sheer chutzpah of the Jefferson-displacing Texas State Board of Education underscores the political impotence of secularists. Similarly, the drift of the U.S. Supreme Court away from separationist positions in recent decades points to something that few reality-based secularists can ignore: while American secularism has many enemies, it has few elected defenders and little access to the corridors of power.

Perhaps this accounts for the hyperbolic claims made by self-identified secular leaders and intellectuals. They reason that although they lack political power, they retain a massive base of sympathizers – 27 million, 29.5 million – who can be mobilized. But the truth is less reassuring: secularism as currently advocated lacks both political power *and* a massive base of sympathizers.[11]

Which could lead a person to wonder whether any iteration of secularism in the United States ever had political clout or significant numbers. Has secularism always been the delusional sad sap of American *isms*? Or has it on occasion flexed its brawny, bare arms in arenas of influence? The answers to these questions will tell us much about our subject's seamier side as well as some nearly forgotten moments in American religious history. It will also take us to the 1960s, the decade when a different possibility for secular identity came into play. What we shall see is that the story of secularism's rise and fall – like everything else about it – is complex, paradoxical, and poorly understood.

American Nonbelief: The Gimpy Zebra

Secularism in America today is usually understood and practiced as either separationism or nonbelief. Some think of it in starkly *political* terms as a strategy that assiduously walls off the government from religion. Others think of it in *theological* terms as an affirmation of the nonexistence of God. Not all separationists are nonbelievers. But nearly all nonbelievers are separationists. This too compounds the confusion. Nonbelievers, such as those already mentioned, are the people who now speak for secularism – by which they really mean atheistic total separationism.

As we try to puzzle out the complicated and virtually unexplored question of the rise of American secularism, we must always bear in mind the difference between its separationist and atheist variants. Each draws upon a distinct intellectual tradition and has its own history. Separationism was a coarse slab of theological intuition that was quarried during the Reformation. Centuries later it was hewn into a coherent policy position by Jefferson and Madison. From there it was set into the metaphorical wall by twentieth-century jurists.

Atheism arrives on the scene a little later. Its first true signs of group life can be detected in the radical Enlightenment of the eighteenth century, especially in France. By the nineteenth century full-blown infidel movements mushroomed in England and, as we are about to see, in the United States. Separationism and atheism, although distinct, have something of an elective affinity. Over the two centuries or so in which they have "shared the air," they have tended to find each other. They clearly coalesced in the activism of figures like the Victorian Charles Bradlaugh. The same coupling occurred – with grim results – in the Soviet Union.

In the United States separationism and atheism have also melded in a distinct way. What must be stressed, however, is that in American history *substantial political gains and popular appeal have accrued only to separationist secularism,* not to atheist secularism. Separationists have scaled stunning political heights and sunk to regrettable moral lows. Nonbelieving secularists, by contrast, have occasionally "made a little noise" and scandalized some squares. For the most part, however, they have remained marginal within American politics and culture.

Nonbelievers today, as we have seen, view things differently. To hear them tell it, their numbers are booming and their agenda is advancing. The truth is strikingly different. American atheist movements, though fancying themselves a lion, are more like the gimpy little zebra crossing the river full of crocs. In terms of both political gains and popular appeal, nonbelievers in the United States have little to show. They are encircled by cunning, swarming Revivalist adversaries who know how to play the atheist card.

True, Barack Obama did give a shout-out to nonbelievers in his inaugural address, but that was highly unusual.[12] American presidents have been at best apathetic and at worst hostile to nonbelievers. And besides, as we shall see, what President Obama gives with one hand, he takes away with the other.

Moving from the executive to the legislative branch, no atheist movement in the United States has ever achieved the type of popular critical mass that would force lawmakers in the House to take notice of its agenda. It is not a coincidence that there is a Congressional Black Caucus, but not an atheist one, nor a secular one, for that matter. There are no self-professed atheists in the House and Senate and this has been the norm for most of American history.[13]

Unfortunately, nonbelief is, and always has been, treated with contempt in the American public square. If politicians play the atheist card, it is because atheist groups are small and anti-atheist sentiment large. We observed this in the 2008 Senate race in North Carolina. There, the campaign of the incumbent Republican senator Elizabeth Dole released a thirty-second attack ad. It accused her challenger, the Democratic state senator Kay Hagan, of accepting campaign funding from the Godless Americans Political Action Committee.[14]

"Kay Hagan took 'Godless' money," it began. "What did Kay Hagan promise in return?" asked the announcer darkly. Then a female voice, accompanying a photo of Hagan, intoned, "There is no God."[15] Hagan responded with an ad of her own, affirming that she "believe[s] in God" and "taught Sunday school."[16] For good measure, she also filed a defamation lawsuit against Dole.[17] Calling an American the A-word is an actionable offense![18]

A glance at current atheist and nonbelief movements in the United States indicates that they do, in fact, make easy targets. They are small and underfunded. They lack the muscle of Washington power players like the NAACP or the Human Rights Campaign. If atheists had organizations with clout like that, politicians would think twice before engaging in infidel baiting. The American political scene is a savannah dominated by apex predators. Atheist advocacy groups can chest-thump all they want, but the gloomy reality is that they are small, slow, vulnerable, and overwhelmed by Revivalist activism.

The numbers, contrary to the claims we have already encountered, bear this out. The Freedom from Religion Foundation claims to have over sixteen thousand members who are "freethinkers," defined by FFRF as agnostics or atheists.[19] In 2003, the erstwhile executive director of the American Humanist Association described his organization as having "6,000-plus" members.[20] As of 2005, the American Atheists – whose founder we shall meet in a moment – reported about twenty-two hundred members.[21]

Secular atheists have never achieved legislative or judicial power, nor have they garnered the numbers that they often boast about. What they have attained, on occasion, is nationwide infamy. Periodically throughout American history, individual atheists have scored significant media attention. By frankly expressing their views on God, the Bible, what have you, they have repeatedly outraged their fellow Americans who are religious (and who are, we hasten to add, too easily outraged on matters religious). The media, for its part, has always been ready to relegate any discussion of nonbelief to "circus sideshow" status.[22]

As a publicity strategy, infamy is not the worst thing in the world. It draws attention to your movement, helps you focus your message, dog-whistles to the base, and musters a few recruits along the way. Today's New Atheists have, to their credit, established a formidable media platform for radical anti-theism. This platform has not, however, translated into any substantive political gains (the argument could be made that things have actually gotten worse for movement secularists since the New Atheists blasted off in 2004). On the other hand, it has done wonders to raise the profile of anti-theism in the United States.

A generation ago, the role of national atheist firebrand was played by Madalyn Murray O'Hair (1919–1995). At first a sensation and eventually a mere spectacle, O'Hair became "American atheism's leading proponent, educator, spokesperson, and symbol."[23] In 1964 *Life* magazine referred to her as "the most hated women in America." O'Hair founded the group American Atheists (significantly, she initially called it the Society of Separationists).[24]

Unlike the New Atheists, she could boast of a few actual political accomplishments. She was the plaintiff in the *Murray v. Curlett* case of 1963, which was folded into the landmark *Abington School District v. Schempp* decision, whereby the Supreme Court ruled that at the start of the day in public schools, Bible reading, even when unaccompanied by comment, was unconstitutional.[25] Following in the tradition of atheist self-aggrandizement, she affirmed that "atheists have the biggest underground movement in America. They are everywhere."[26] In 1969, O'Hair claimed that her advocacy served 74 million Americans.[27]

Her antics, such as proclaiming Thursday as the Sabbath for the godless in America, were legion.[28] Yet she would eventually run the small national movement into the ground. O'Hair mired American Atheists in factionalism and chronic controversy, not to mention embezzlement scandals.[29] The story ended terribly, with O'Hair, her son, and her daughter-in-law murdered by a former employee.[30]

In terms of media scrutiny and buzz-worthy infamy, the predecessor to O'Hair was the inimitable Robert Ingersoll (1833–1899).[31] Although exceptional figures like Ingersoll, O'Hair, and the New Atheists have never raked in millions of new members, they have drawn sizable crowds and curried journalistic interest.[32] Between 1869 and 1899, Ingersoll "toured the nation almost continuously, delivering six or seven lectures a week to sold-out houses."[33]

The Pagan Prophet, as he was known, was by far the most famous of the nineteenth-century freethinkers.[34] *Freethought* is a term that encompasses an extraordinarily wide variety of countercultural positions and attitudes. These include but are not limited to religious skepticism, scientific rationalism, agnosticism, socialism, and full-blown atheism. Another term that was, interestingly, used as a synonym for *freethought* and *secularism* in the nineteenth century was *liberalism.* Also, the freethought movement attracted *religious* liberals. They might include members of Unitarian and Universalist churches, those who dissented from their own Protestant traditions, Reform Jews, and idiosyncratic Catholics.

The golden age of American freethought spanned the post–Civil War period up to the outbreak of World War I in 1914.[35] Ingersoll was at the center of this ferment. He had crossed paths with all of the major "infidel" individuals and organizations of the era and pollinated them with his ideas. As president of the National Liberal League, which was founded in 1876, he tried to rein in the crippling factionalism that so often afflicts organizations of freethinkers and nonbelievers.[36]

Ingersoll eventually left the group because it had become too radical.[37] He then became president of its offshoot, the American Secular Union, in 1884. Here he oversaw the continuing build-out of what may be the single most important platform in the history of American freethought.[38]

In 1872, the freethought magazine *The Index* published "Nine Demands of Liberalism."[39] In the words of one scholar, this "platform was repeated in the pages of every freethought periodical and has never been discarded as the central program of the freethinking secularist movement."[40] Among the Nine Demands are desiderata such as eliminating the tax exemption from "churches and other ecclesiastical property"; removing chaplains from all government institutions, including the armed forces; and discontinuing "all public appropriations for sectarian educational and charitable institutions."[41]

Ingersoll had thought long and hard about secularism as a philosophical concept. His definition was clearly influenced by the ethical concerns of George Holyoake's work (the two were friends and correspondents).[42] For Ingersoll, secularism was – well, it was actually a bunch of things: "the religion of humanity; it embraces the affairs of this world; it is interested in everything that touches the welfare of a sentient being . . . it is a declaration of intellectual independence; it means that the pew is superior to the pulpit . . . it is a protest against ecclesiastical tyranny, against being the serf, subject, or slave of any phantom or of the priest of any phantom; it is a protest against the wasting of this life for the sake of one that we know not of; it proposes to let the gods take care of themselves. It is another name for common sense."[43]

Though a formidable orator and a likable character, Ingersoll is largely forgotten today.[44] Freethought never snowballed into a mass movement in the United States. Indeed, the term has floated out of our national lexicon. As for the three crucial elements of the Nine Demands cited here, they remain unheeded.

American atheist secularism has, sporadically, attained notoriety in our nation's history. It has yet to register political accomplishments or widespread popularity. If we want to discern signs of secularism's political or demographic rise, we must look elsewhere.

Separationist Secularism in the Nineteenth Century

Although atheism has never achieved a foothold in American society, separationist secularism, albeit not of the "total" separationist variety, has wielded impressive power and influence at least three times in American politics.

American separationist secularism took legislative shape for the first time in the late eighteenth century. Its champions were anything but atheists. The early separationist vanguard combined evangelical religious dissenters (with some Baptists leading the charge), deist intellectuals, and people who were difficult to categorize as to religion.

That two of the architects of the secular vision, Jefferson and Madison, lived in the White House for sixteen consecutive years speaks volumes about the political prominence of separationism in this country. We have already seen how Thomas Jefferson made the essential case for keeping the church out of governmental affairs. His activism on behalf of this radical position as well as his own highly idiosyncratic religious views – Jefferson once referred to himself as "a sect unto myself" – did not always win him admirers.[45] In fact, it won him scads of detractors, many of whom were eager to tar him as, yes, an atheist!

Running for president as a Republican in 1800, the great Enlightenment thinker (and doer) was viciously attacked as a nonbeliever by the rival Federalist Party. He was lambasted as a godless man, a despiser of religion, a supporter of French Jacobin butchery (that is, the ideas of the French Revolution), a desecrator of the Bible, and, in sum, "a dangerous, demoralizing infidel."[46] The (willful) confusion between separationism and atheism stretches back some two centuries.

Republicans responded with heat, coming to Jefferson's aid and supporting the policy prescription most associated with his name. His defenders argued in favor of separation. What is important for our purposes is that they did so along *theological* lines. For them separationism had absolutely nothing to do with atheism (it is highly doubtful, by the way, that there were more than a handful of real live atheists in America at this time).[47] As one Republican exclaimed, "It is your duty, as Christians, to maintain the purity and independence of the church, to keep religion separate from politics, to prevent an union [*sic*] between the church and state, and to protect your clergy from temptation, corruption, and reproach."[48]

There was no shortage of pious, God-fearing nineteenth-century clerics and theologians touting the godliness of walling off. This points to an association, long forgotten, between Christian dissenters and the push for a religion-free government. There exists a rich, but now obscured, tradition in America of *Christian* secularism (understood as separationism). Secular Republicans of the era held, astonishingly, that "America could never become in spirit a true Christian nation until complete church-state separation had been secured."[49]

Readers might be intrigued by the phrase "secular Republicans." Today such a species is quite rare, but this was not the case in the nineteenth century. To properly understand that phenomenon, we must first look back to the 1840s. In this decade the separationist mindset, now forty years removed from the booming cadences of Jefferson's Danbury letter, took on a new ideological attribute: rabid anti-Catholicism.

With each passing decade of the nineteenth century, more and more Catholics immigrated to the United States. Between 1810 and 1860, the Catholic population of America swelled from seventy-five thousand to three million, representing nearly 10 percent of the population.[50] Old World animosities die slowly. The new arrivals encountered considerable discrimination. As the scholar Philip Hamburger observes, "Protestants assumed that the Catholic Church, because of its apparent threat to intellectual independence, was all the more likely to obtain political power and revive a medieval intolerance. In response to this combination of liberal and more traditional fears, Protestants would eventually elevate separation of church and state as an American ideal."[51]

There was a pervasive unofficial establishment of Protestantism in nineteenth-century America; Jews and Catholics were the ones most likely to run afoul of its unwritten laws. For Catholics this first occurred during the "Bible Wars," one of the grimmest and least scrutinized chapters in American religious history.

The Bible Wars first broke out in the 1840s. The flashpoints were urban areas where Catholic and Protestant children attended "common schools," early versions of our current public schools. Some of the most notable clashes about school policies occurred in New York, Newark, Philadelphia, and Cincinnati. It was during these ructions that the concept of separation "first attracted wide support and even national attention."[52]

The debates typically centered on Catholics' misgivings about the use of the Bible for prayer and instruction at school. In this period, the scriptures were invoked in opening exercises and also used as a sort of textbook, or "school reader."[53] But whose scriptures to use? The majority Protestant culture naturally favored their King James Version. The minority Catholics naturally preferred Catholic translations, such as the Douay.[54]

Catholics in the 1840s tried a variety of strategies to avoid subjecting their children to the Protestant King James Bible. They proposed letting Catholic students use the Douay.[55] Sometimes they requested that Bible readings be removed altogether from the public schools.[56] This request was usually but not always denied.

After much conflict leading to more than a few bloody urban riots (a particularly frightful and deadly one occurred in Philadelphia in 1844), Catholics opted for a different strategy.[57] Frustrated with the xenophobic Protestant establishment, they eventually aimed to create their own private, or diocesan, institutions in accord with the motto "every Catholic child in a Catholic school."[58] This would set the stage for a new nationwide showdown in the 1870s.

Why couldn't any of the Catholics' demands be met, especially the request to simply rid the schools of the Bible altogether? Over and over again, Protestant officials concluded that the Protestant King James Bible was "nonsectarian" or "nondenominational." As a "neutral" text, it was deemed appropriate for all public school pupils. Even if the students were not Protestant, it was reasoned, the text would be suitable and salutary for them. Through it all, many Protestants cast themselves as defenders of the idea of separation.

To a modern observer the double standard and hypocrisy are truly staggering. These schools were promulgating *Protestant* scriptures and *Protestant* prayer, all the while insisting that they were acting in the name of church-state separation.[59] Historians themselves often seem perplexed that "American public opinion saw no contradiction in simultaneously embracing the concept of separationism and the practice of public prayer [in schools]."[60] It is very difficult to understand how Protestants could have remained oblivious to this massive trespass on the separation of church and state. Perhaps posterity will find similar blind spots in our generation.

Then again, there is ample reason to suppose that this was no blind spot, but instead naked, self-conscious anti-Catholics bigotry. Many of the separationists in this debate attacked Catholics as "papists" and "foreigners" who allegedly sought to sell out America to Rome.[61] The secular vision, as we have seen, is Protestant to the core, and here one of its less salutary impulses came to the surface.

These midcentury debates were the prelude to the infamous Blaine amendments of a few decades later. One consequence of the Bible Wars, as we have seen, was that frustrated Catholic Americans began to establish their own schools. In a few cases the increasing political power of Catholics garnered indirect public funding for these schools in places like New York and Wisconsin.[62] Enter James G. Blaine, a congressman from Maine and a former Speaker of the House, with an eye on a presidential run. (It is fascinating to note that the aforementioned freethought icon, Robert Ingersoll, was a close friend and supporter of Blaine and once called him "the grandest combination of heart, conscience, and brain beneath the flag.")[63]

The Republican understood that the "scandal" of state funding of those Catholic schools would play well among the nation's not inconsiderable legions of anti-Catholic voters.[64] In 1875 Blaine introduced legislation in Congress that would prevent Catholic "sectarian" schools from receiving any such funding: "No states shall make any law respecting an establishment of religion, or prohibiting the free exercise thereof; and no money raised by taxation in any State for the support of public schools, or derived from any public fund therefore, nor any public lands devoted thereto, shall ever be under the control of any religious sect; nor shall any money so raised or lands so devoted be divided between religious sects or denominations."[65]

Blaine's quest for the ratification of his amendment fell slightly short and the proposed Sixteenth Amendment to the Constitution never came to pass. Yet it was the afterlife of Blaine that made the greatest impact. By 1890, twenty-nine states in the Union had banned any sort of state funding for sectarian schools.[66] Today, proponents of school voucher plans in the United States find the existence of Blaine amendments in state constitutions, thirty-six in all, to be a huge hindrance (separationists, for their part, might use these amendments to their advantage).[67]

We need not linger over the complex backstory regarding the amendments nor their implications for states' rights. For purposes of understanding the rise of secularism, a few facts are noteworthy. The first is that evangelical Protestants were staunch supporters of the Blaine amendments. What makes this all very confusing is that these evangelicals were speaking in the name of pseudo-separationism: they permitted Protestant Bibles and prayers in public schools while excluding all others! Even more striking is the presidential legitimacy granted to the separationist worldview. The remarks of President Grant in this chapter's epigraph indicate that separationism, once again, had the blessing of the executive branch.

This came at a moral cost. The Blaine episode points to a dark interlude in the history of secularism. The champions of separationism are not cast in one positive mold: they can be as morally serious as Jefferson or as unsavory as the anti-Catholic hatemongers of yore. Today's secularists reflexively tend to see themselves as endowed with high moral purpose, but the xenophobic nativism of an earlier era reveals a checkered past.

Secularish: The Sixties

In recounting thus far the history of separationist secularism as a politically formidable movement supported by a sizable constituency, two periods have come to our attention. The first is the glory days of revolutionary and post-revolutionary America. The next is the second half of the nineteenth century, when separationism merged successfully, albeit disturbingly, with anti-Catholic social movements. Presidents, senators, members of Congress, broad popular support – separationist secularism, for better or for worse, had a bunch of that in the nation's first hundred years.

The third historical peak of secularism is one we have already encountered and will continue to scrutinize. This is the period from the 1940s through 1970s, when separationism became the preferred secular policy of the U.S. Supreme Court. We have reviewed some of the landmark decisions of that era, during which the wall of separation stood higher and sturdier than at any point before or after.

But did judicial separationism have widespread popular appeal in this period as well? This is a crucial question and we will investigate it in the next chapter. Before doing that we need to look a bit more carefully at the anything-goes 1960s, perhaps the signature decade in the history of American secularism. There, we can find popular support for something that can be called neither separationism per se nor atheism. Rather, tens of millions of Americans in that decade looked at the world through more softly tinted secular lenses and adopted views that we shall refer to as secularish.

This is certainly not to say that everything about the 1960s was good, or just, or worthy of our respect today. Nor is it to say that secularism was a particularly coherent or widely understood concept back then. Yet what we glimpse in this decade are radically new ways of thinking about politics and faith. If secularists want a crack at those aforementioned 27 million or 29.5 million recruits, mentioned (and imagined) earlier, they had better take a good look at the new worldviews that emerged half a century ago.

As we just observed, the 1960s was a sterling decade for separationism as espoused by the government. Things got off to a rousing secular start in September 1960 when Senator John F. Kennedy, campaigning for the presidency, boldly told a group of suspicious Protestant clergy in Texas that he believed "in an America where the separation of church and state is absolute."[68] Those pastors of the Greater Houston Ministerial Association who were vetting JFK had fears about Catholics that had not changed much from the days of the Bible Wars and the Blaine amendments.

Kennedy's dictum expressed the consensus opinion of that time. His Protestant listeners – who envisioned papal flags streaming aloft on Pennsylvania Avenue and the Swiss Guard standing watch over the West Wing – *wanted* to hear the Catholic democratic nominee say precisely what he said. Only many years later would Protestant anti-separationism (in the form of Revivalism) assume its current shape, rationale, and audience. Part of this has to do with the fact that white conservative Protestants in the 1960s were focused on the church, not the state.

As for other Protestants, they were thinking about both, but in ways that exemplify the innovative spirit of the times. Take the central political force of the 1960s, the civil rights movement. It is hard to get a grip on what its leader, Martin Luther King Jr., made of Supreme Court decisions prohibiting school prayer or release time for religious instruction. King devoted very little attention to the subject of relations between church and state. Yet one of his rare remarks on this issue is intriguing. "The church," he said in a sermon, "must be reminded that it is not the master or the servant of the state, but rather the conscience of the state." The church, continued King, "must be the guide and the critic of the state, and never its tool. If the church does not recapture its prophetic zeal, it will become an irrelevant social club without moral or spiritual authority."[69]

Notice that King invokes Luther's old intuition about the necessity of partitioning government power from religion. Yet the great civil rights crusader was not a total separationist. He posits an important place for religion in public life. His suggestion that the church play the role of "moral guide" is not, in and of itself, objectionable. A secular state could, in principle, listen to religion. The trick is to avoid listening to only *one type* of religion (a frequent criticism, we shall see, leveled at the recent Bush administration). As long as the viewpoints of numerous faiths are considered, we could parse King's approach as a softer secular alternative to total separation. This is known as "accommodationism" or "nonpreferentialism" and there is more to say about it later.

Which brings us to the domain of the secularish. This book's introduction mentioned that secular forms of governance have secondary, or bonus, effects. It was in the 1960s and the succeeding decades that some of these effects came to fruition. The secularish 1960s witnessed intense questioning of religious dogma and orthodoxy. As two historians of this decade note, "Between the mid-1950s and the mid-1980s, over a third of all Americans left the denomination in which they had been raised."[70] Others stayed in their traditions but engaged in critical reassessments, which were often brutal in their frankness.

Consider the Jewish theologian Richard Rubenstein, who created a minor outrage with his 1966 book *After Auschwitz*. Combining rapid-fire prose with deep existential brooding, the author took as his theme the idea that "we live in the time of the death of God."[71] For Rubenstein, God died in the Nazi killing camps, and for this reason He "is totally unavailable as a source of meaning and value."[72]

Lest this statement shock his audience not quite enough, he refers to God as "the Holy Nothingness" and laments that "we are alone in a silent unfeeling cosmos."[73] Oh, let's not forget to mention that Rubenstein was a believer, an ordained rabbi, and associated with the mostly Protestant circle known as the death of God theologians (itself a remarkable secularish cohort).[74]

In addition to such critiques of one's own religion, the 1960s witnessed an unprecedented curiosity regarding other faiths. Even the Second Vatican Council of 1962–1965, in its own way, validated a type of secularish analysis. The nineteenth section of its pastoral constitution, *Gaudium et spes*, reads, "Atheism must be accounted among the most serious problems of this age, and is deserving of closer examination."[75] Whereas a student of church history might expect the document to anathematize the atheist, no such condemnations are forthcoming. Rather, what follows is a sober, fair, and introspective analysis of the significance of nonbelief for Catholic thought, including statements like this: "Conscious of how weighty are the questions which atheism raises, and motivated by love for all men, she [the Church] believes these questions ought to be examined seriously and more profoundly."[76] Not exactly a teary embrace of nonbelief. Nonetheless, the willingness to look at the problem, calmly and without rancor, is impressive.

Gaudium et spes closes its discussion of atheists as follows: "While rejecting atheism, root and branch, the Church sincerely professes that all men, believers and unbelievers alike, ought to work for the rightful betterment of this world in which all alike live. Such an ideal cannot be realized, however, apart from sincere and prudent dialogue . . . She courteously invites atheists to examine the gospel of Christ with an open mind."[77]

In Revivalist Catholic precincts nowadays, the liberalizing Vatican II initiatives are seen as an abomination. Yet that brief window of time reveals a type of inquisitiveness on the part of the church, a desire to understand the other. *Gaudium et spes* legitimated and stimulated earnest Catholic engagement with the church's nonbelieving friends.

Thus a Catholic theologian such as Karl Rahner could see atheism as a false interpretation that was nevertheless "a genuine experience of deepest existence."[78] For a good decade or so after the breakthrough that was *Gaudium et spes,* super-groovy Catholic theologians in America and elsewhere would routinely spin off sentences like this one: "The questions that atheism poses to the believer today can help him better understand that faith in God involves faith in man."[79]

So far we have identified this golden age as an era of religious self-critique and religious curiosity. To this we wish to add another secularish quality, anti-authoritarianism. A major theme of the women's liberation movement was the attack on patriarchy, and this intervention occurred in religious institutions as well. The critique of male supremacy came "from within" and could be forged by nuns in the Catholic Church, or the women's caucus of the National Council of Churches, or female Baptists.[80] One feminist from the 1970s put it this way: "While women do comprise the large majority of active church members and are the sustaining force in almost every congregation, they have virtually no power within its structure . . . they are viewed as helpers for the men, with their only real talent seen to be in aesthetic matters or in working with children."[81] Another writer rails against the "male-dominated and male-oriented churches."[82]

This anti-authoritarian streak was most often directed at the powers that be. In a decade of antiwar activism, those powers could be construed as the government. But they could also be construed as the Protestant establishment.

A study of the pornography business in the sixties, for example, reveals the large role played by Jews as both actors and producers. One observer notes that "Jewish involvement in the X-rated industry can be seen as a proverbial two fingers to the entire WASP establishment in America. Some porn stars viewed themselves as frontline fighters in the spiritual battle between Christian America and secular humanism."[83] Strange as it may sound, the secularish sixties unleashed similar critical impulses among pornographers and feminist theologians alike!

The roiling 1960s stands as the high-water mark of American separationism. With this strategy of secular governance in place, citizens were free to create and explore a secularish worldview. Its fundamental attributes were as follows: (1) a critical self-questioning of religious dogma whose ultimate destination is not necessarily nonbelief, but deeper comprehension of the divine; (2) a curiosity about other faiths and what they may teach us; and (3) an anti-authoritarian streak that challenges religious orthodoxy, dogma, and power.

Of course, while the separationist and secularish bonfires were raging in the 1960s, one group watched aghast. Once they started channeling their anger, American secularism would be rocked to its foundations.

7

The Fall of American Secularism

> They [the Supreme Court], as an elite, thus force their will on the majority, even though their ruling [in *Roe v. Wade*] was arbitrary both legally and medically. Thus law and the courts became the vehicle for forcing a totally secular concept on the population.
>
> — FRANCIS SCHAEFFER, *A Christian Manifesto*

AT FIRST GLANCE there is one devastatingly all-encompassing explanation for the sorry state of American secularism today: the conservative Christian Revival of the late twentieth century. Over the past few decades, the Christian Right has pulverized secularism. Simply *pulverized* it! While secularists were perusing Philip Roth or listening to jazz or downloading porn or God knows what, many of their old haunts and love shacks were being systematically overrun.

There is no shortage of metrics demonstrating how far we have fallen from separationism's midcentury peak. The Warren and Burger Courts morphed into the Rehnquist and Roberts Courts. Their majorities have steadily undone decades of First Amendment accomplishments of the 1940s through the 1970s. The public school has mutated into a place where "intelligent design" advocates hunt down with Darwinian ferocity any textbook that presents the theory of evolution as valuable scientific information. Presidential elections, events that in the past were marked by restraint where religion was concerned, metastasized into veritable Christ-fests in 2000, 2004, and 2008. The few abortion clinics that are left standing outside major cities have devolved into high-security militarized zones.

Over more than forty years of struggle, the Christian Right has been better organized, disciplined, funded, and focused than its secular adversaries. Not that these adversaries have put up much of a fight. In fact, it might not be an exaggeration to claim that, until recently, secularism *didn't even know it had a fight on its hands*. The boast of Ralph Reed, the legendary Christian Right mega-operative, comes to mind: "You don't know it's over until you're in a body bag. You don't know until election night."[1]

Reed, who was later embroiled in the Abramoff scandal, ran evangelical and fundamentalist "stealth candidates" in elections, major and minor, across America.[2] The executive director of the Christian Coalition in the early 1990s typically fronted office seekers who could be called nondescript. By design, they were "young, inexperienced Republicans" lacking any sort of "prior record" – all the better to mask the fact that they were Bible thumpers who had accepted Christ into their hearts and were hell-bent on "taking back America."[3] Thus Reed seeded a formidable Revivalist grass-roots network. It sprouted everywhere, from the PTA to the statehouse. Secularists tended to learn about his body of work on the morning after election day, if only because they *were* the body of his work.

The story of the rise of the Christian Right has been told many times and we need not reprise it at length, except to identify the trigger that unleashed it: the separationist and secularish 1960s. That decade scandalized conservative Christians and snapped them to attention. For them it was like a petri dish full of germs that transmogrified into all manner of abominations during the 1970s and beyond. That period witnessed the hatching of several frightful secular beasts: the removal of prayer from public schools, the drug culture, feminism, the gay rights movement, a massive breakout of the previously furtive porn industry, the sexual revolution, and, of course, the legalization of abortion.

What is intriguing about the Christian Right is the influence its intellectuals have on its activists. The former surveyed the catastrophe, identified the root causes, and selected the targets. The latter locked and loaded. These Revivalists have produced many cultural critics over the past few decades, but in terms of this book's concerns, perhaps none was more important than the evangelical theologian Francis Schaeffer (1912–1984).[4] An eccentric and a polymath, Schaeffer sounded an agenda-setting alarm in the 1970s and 1980s. In many ways, he wrote the manual for the conservative evangelical counterinsurgency, which then radically altered the shape of American politics.[5]

He was quite a character, Mr. Schaeffer was. Fond of wearing knickers, exuding a guru vibe, and citing Bob Dylan's lyrics, he set up his base in Switzerland. From there he proceeded to influence an entire generation of American evangelicals. Schaeffer was a far more interesting and thoughtful individual than his countless (lawful) extremist acolytes (who include figures like Tim LaHaye, author of the Left Behind series). But he was an extremist nonetheless.

Schaeffer would gravel the intellectual path for conservative Christianity's march on secularism. In 1981, anticipating the assault on separationism chronicled here, he described the United States as being "founded upon a Christian consensus."[6] "To have suggested," he demurred, "the state [be] separated from religion and religious influence would have amazed the Founding Fathers."[7]

Schaeffer was one of the first to set secularism, or what he also referred to as "humanism" or "secular humanism," squarely within Revivalist cross hairs. He noted, "Ironically, *it is the humanist religion* which the government and courts in the United States favor over all others!"[8] Fuming over a challenge to a creationist curriculum, he raised this objection: "The ACLU is acting as the arm of the humanist consensus to force its view on the *majority* of the Arkansas state officials."[9] For Schaeffer, secular humanists were a scourge. They were "determined to beat to death the knowledge of God and the knowledge that God has not been silent, but has spoken in the Bible and through Christ—and they have been determined to do this even though the death of values has come with the death of that knowledge."[10]

With intellectuals like Schaeffer outlining the challenge, the Revivalists began their slow, steady inundation of the political sphere. As Jerry Falwell lamented during preparations for the counterattack, "For thirty years, the Bible-believing Christians of America have been largely absent from the executive, legislative, and judicial branches of both federal and local government."[11] Having weathered the storm of the sixties, it was time to get back in the game.

Following the lead of conservative Catholics, evangelicals eventually got around to protesting the 1973 *Roe v. Wade* decision legalizing abortion. Although it is not widely known, many conservative Protestants were initially at peace with, or apathetic to, the *Roe* decision. It was traditionalist Catholics, those first responders to culture war calamities, who sounded the initial alarm.[12] Roused from a long electoral slumber by the candidacy of Jimmy Carter, many white conservative evangelicals cast their votes for a Bible-believin' Democrat in 1976.[13]

Carter, a liberal evangelical, soon disappointed them (and so many others). So in 1980, conservative Christians threw their support to Ronald Reagan. Galvanized by leaders like Jerry Falwell and his Moral Majority, and later, Pat Robertson and his Christian Coalition, Revivalists became a mainstay of the GOP. Their thirty-year association with the party has had its ups (the campaigns of George W. Bush in 2000 and 2004) and its downs (the 2008 presidential run of John McCain, at least until Sarah Palin came aboard). But in the main, it remains solid and healthy up to this very day. The road to any Republican presidential nomination still must pass through evangelical America.

Secularism quite simply had no answer for any of this. It wasn't even asking questions. Or more precisely, there weren't enough secularists around to ask questions. What happened to secularism? In a phrase, the Christian Right.

Still, there were serious *internal* deficiencies that facilitated the Revivalist conquest. Some of these had to do with vulnerabilities and weaknesses unique to American secularism. Even as it was experiencing its greatest triumphs, secularism's tactics and organizational foundations were exceedingly flimsy. And still other shortcomings exposed a deep structural flaw in its fragile Reformation-and-Enlightenment-forged heart.

The Judicial Strategy

A collection of American secular action figures would surely feature the likenesses not only of Jefferson and Madison, but also Justices

Hugo Black and William Douglas. It seems like only yesterday that the Supreme Court was the home field of separationist secularism. It was, after all, because of the nation's highest court that secularism crested from the 1940s to the 1970s.

Revivalists concur. To hear them tell it, the Court in that era was the sinister citadel of secularism. It was an institution hell-bent on scrubbing the public sphere of all signifiers of faith. Its diabolical objective was to de-Christianize our great nation. One observer of the religious Right points out that "no conservative or evangelical discussion of the Warren Court or its rulings is complete without some variant of the phrases 'judicial activism,' 'judicial tyranny,' or 'judicial dictatorship.'"[14]

Yet evangelicals and fundamentalists clearly understood that the Court is not, so to speak, constitutionally secular. With every retirement and every new appointment, its ideological texture is likely to change. The religious Right has spent decades plotting how to bring about this change and thereby recapture the judicial branch. They may be well on their way to accomplishing this as they flood the offices of the federal government with conservative Christian lawyers. It is no coincidence that the law school at Jerry Falwell's Liberty College contains an exact 4,395-square-foot replica of the U.S. Supreme Court.[15]

In the aftermath of a scandal at the Department of Justice in 2007, Americans learned about another Christian law school whose graduates were working on behalf of a "biblical worldview." At the epicenter of that flair-up was the attorney Monica Goodling, a graduate of Pat Robertson's Regent University Law School. As senior counsel to Attorney General Alberto R. Gonzalez, Ms. Goodling oversaw the firing of Department of Justice lawyers for apparently partisan reasons.[16] Recently, Louisiana College opened its own law school. Its charter: "To address judicial activism and the erosion of religious liberty." To that end it "will produce attorneys who will be Christian advocates in everyday life and practice as well as the realm of politics."[17]

Through slow, patient activism, the Christian Right has made great political strides. Like-minded operatives have been well trained and deployed throughout the lower courts. With judges like John Roberts, Sam Alito, and Clarence Thomas in place, the foundation has been laid for an eventual tearing down of the wall of separation as we know it.[18] For many observers, it's not a question of *if* this happens, but *when.*

Indeed, the recent decisions of the Supreme Court have not been kind to secularism. A case like *Arizona School Tuition Organization v. Winn et al.* in 2011 contained a lot to put secularists on edge. For example, the decision made it increasingly difficult for citizens to bring Establishment Clause cases to the Court's attention. The refusal to grant "standing" to those seeking redress about the religion clauses was truly stunning; few of the major legal milestones in American secular history would have been achieved if this new standard applied half a century ago.

In her first dissent – and a crackling one at that – Associate Justice Elena Kagan warned that the decision "offers a roadmap – more truly, just a one-step instruction – to any government that wishes to insulate its financing of religious activity from legal challenge . . . No taxpayer will have standing to object. However blatantly the government may violate the Establishment Clause, taxpayers cannot gain access to the federal courts."[19] We spoke earlier of monumental desecularizing shifts in American politics and society. Consider the last time the Court addressed the issue, which was in 1968; by an 8–1 majority it affirmed that citizens did indeed have standing to bring such litigation.[20]

A few hard truths about the Warren and Burger era need to be noted to help make sense of this irony: even as secularism rose, it was falling. First, many of the decisions that the midcentury Court reached about separation of church and state were unpopular with the American public. Take, for example, the *Engel v. Vitale* decision of 1962, which struck down state-sponsored prayer in New York public schools. According to a Gallup poll, 79 percent of Americans at that time approved of such religious exercises.[21]

A year later, in *Abington School District v. Schempp,* the Court held that required Bible reading and recitation of the Lord's Prayer in public schools were unconstitutional. As for that ruling, "70% of Americans disapproved of the Court's decision."[22] Moving, for purposes of comparison, into recent times, a CNN/USA Today/Gallup survey in 2005 found 76 percent of Americans in favor of the display of the Ten Commandments in the Texas Capitol.[23] In the wake of a lawsuit by Michael Newdow in 2004, Gallup similarly discovered that 91 percent of Americans wanted to keep the words "under God" in the Pledge of Allegiance.[24]

Secularism's triumphs, then, were not popular or even legislative in nature, but *judicial*. We could refer to this as separationist secularism's "judicial strategy." The legal scholar Noah Feldman makes the telling observation that secularists at midcentury turned to the courts once they realized "what they could not win from legislative majorities."[25]

Pursuing a judicial strategy unattached to any legislative plan or grass-roots organization is a tactic that has served minorities well. What better way for a small group to pursue its unpopular agenda than by convincing the courts to make that agenda the law of the land? Yet as awesome as the results of a judicial strategy unconnected to a legislative agenda may be, there are risks. Namely, that times change, opinions change, and, most important, the ideological drift of the Court changes.[26]

Secularism prevailed in the judiciary but not in the legislative branch. Secularism won in the courts, but it never won hearts and minds. Many Americans felt that Washington, DC, had *imposed* secularism upon them. It took away your child's right to pray in school, at graduation, and at high school football games. It removed the cross from in front of the post office and told you how you could discuss God at work.

Outside the courts, few were won over by the allures of separationism. Rightly or wrongly, many felt they lived in a secular apartheid state. The evangelist Billy Graham got to the gist of the matter when, after the *Schempp* ruling, he complained, "Why should the majority be so severely penalized by the protests of a handful?"[27]

Who Was Secular?

Surely more than "a handful" of Americans were involved in secular separationism's mild golden age of the 1940s to the 1970s. Yet Graham was quite correct in pointing out that, comparatively speaking, secular separationism was a small, boutique-style *ism* (the secularish mentality was far more widespread). Separationism achieved its judicial milestones *in spite of* its tiny size, though this would become a key factor in its epic collapse.

If we were to select an all-star team of twentieth-century warriors who have fought on separationism's behalf, our roster would include (1) the Supreme Court in its glorious midcentury run, from the Warren court (1953–1969) to the Burger court (1969–1986); (2) the Democratic Party, starting with Franklin Delano Roosevelt and ending with the Obama administration; and (3) the numerous small pressure groups that pursued the aforementioned "judicial strategy." Those in the last category formed what one scholar calls "the separationist advocacy coalition."[28] Here, ladies and gentlemen, are the basic ingredients of the mid-twentieth-century secular sausage.

But where are they now? The Supreme Court has moved away from the old separationist orthodoxy. The shift was symbolically marked by Justice William Rehnquist's dissent in the *Wallace v. Jaffree* case of 1985. There, the soon-to-be chief justice would pronounce the words that would trigger the steady demise of the old order: "The 'wall of separation between church and state,'" he avowed, "is a metaphor based on bad history, a metaphor which has proved useless as a guide to judging. It should be frankly and explicitly abandoned."[29]

The Democratic Party, whose (often defeated) presidential candidates wore secularism on their sleeve, also dropped out of the separation business somewhere around the midterm elections of 2006. The party of Kennedy ultimately learned a lesson from the catastrophically failed campaigns of Walter Mondale, Michael Dukakis, and John Kerry. The Obama administration, we shall see, has offered a full-blown substitute for separationism.

Which brings us to the third pillar of midcentury secularism. This would be the separationist advocacy coalition. It was a diverse cohort comprising some ninety groups, with about a dozen or so doing the heavy lifting.[30] It was this coalition that pursued the judicial strategy discussed earlier. That is to say, it engaged in First Amendment litigation, plotted legal challenges, monitored developments around the country, and filed amicus curiae briefs.

The most prominent players included the American Civil Liberties Union and Americans United for Separation of Church and State (originally known in 1947 as Protestants and Other Americans United for Separation of Church and State; the name change occurred in 1972). In addition, there was a formidable Jewish presence consisting of major organizations such as the Anti-Defamation League, the American Jewish Committee, and the American Jewish Congress. Down the line, the most famous Jewish separationist of all, Leo Pfeffer, would leave the fold of these groups and start his own, called PEARL (Committee on Public Education and Religious Liberty), in the 1960s.[31] There was also the much-ballyhooed 1980s start-up, People for the American Way.

Jews in the separationist advocacy coalition were joined at first by a not inconsiderable number of Protestants. The residue of nineteenth-century anti-Catholicism still motivated some to defend the wall of separation. Others remained loyal to the dissenting tradition that led their ancestors to lock arms with Thomas Jefferson. One group, the Baptist Joint Committee on Public Affairs, was (and still is) a stalwart presence in church-state adjudication.[32]

Strange as it may sound, the mammoth Southern Baptist Convention formerly supported separationism. The SBC (which today boasts sixteen million members) is now positioned firmly in the Revivalist camp.[33] It broke away from the Baptist Joint Committee in early 1991.[34] Overall, however, Baptist participation in secular causes is far less than it used to be. Also involved in the Protestant separationist coalition was the National Council of Churches of Christ as well as groups of Scottish Rite Masons.[35]

Other sources of separationist energy emerged in controversies surrounding the public schools. Groups like the American Federation of Teachers, the National Education Association, and the Horace Mann League resisted the idea of state funding for parochial educational institutions. They did so usually out of a fear that the "public 'pie' for education" was limited in size and they would lose funding to their religious competitors.[36]

As we ponder this list of strange separationist bedfellows, two stark facts stand out. The first is that most of the inductees are no longer in the business of separating church from state. The second is that most of these groups *were never in the business of separating church and state to begin with.* With the exception of Americans United, most of these groups defended the secular ideal of separation on the backstroke – their primary concerns or interests lay elsewhere. Secularism's most effective and heroic defenders have rarely been those whose central mission is the defense of secularism.

Take, for example, minority religious groups such as the Jews. They were driven to secularism by an old fear: discrimination at the hands of a Christian establishment. While Jews had political reasons for endorsing separationism, Baptists had profound theological motivations for doing so. These harked back to the secular vision of Roger Williams and Martin Luther. Public school teachers were in the coalition not out of a love for separationism per se, but for professional and pedagogical reasons. As for the American Civil Liberties Union, its central mission is civil liberties, not secularism. American secularism is one of the few *isms* out there that arose on the backs of individuals and groups devoted to *other isms*.

All praise be upon all of these groups. But, with few exceptions, their support for the principle of separation was a means to an end. Let us put this a little differently: Jews and Baptists would not describe themselves as secularists, but rather as secular Jews and secular Baptists, respectively. Ms. Paulicelli, the schoolteacher down at PS 139 in Brooklyn, would have described herself as a teacher, maybe a Catholic teacher, or plausibly, when asked about politics, a secular Catholic schoolteacher. Secularists were not secularists in the way that Christians were Christians; being secular wasn't an all-consuming central identity. Herein lies both secularism's poison and its cure.

Secularism's "We" Problem

In trying to explain the swift disintegration of American secularism in the final decades of the twentieth century, we have looked at its overemphasis on judicial reform and its lack of grass-roots appeal. Yet there might also be a more abstract, more overarching, more pre-twentieth-century explanation for secularism's plight.

Simply put, secularism has a "we" problem. Secularists don't do "we." They don't understand "we," and another "we" seems to have figured that out. Lawful Revivalists, teeming with numbers, have learned how to swarm the secular state through perfectly democratic maneuvering. When Revivalists vote in a bloc, secularists are positively dumbstruck. When Revivalists train their own to become judges, legislators, and public servants, secularists are bamboozled.

This weakness may have its origin in certain features of the basic package that reach back to the sixteenth and seventeenth centuries. The Protestant architects of secularism conceived of the world in terms of *individuals.* It was the solitary human being, not the community, who composed the fundamental unit of political analysis and moral redemption. The architects were deeply suspicious and mistrustful of religious majorities. All sought to protect the individual conscience from the tyrannical impulses of such a majority. The Enlightenment architects went a step further. They endeavored to safeguard not only the individual, but also the *state* from these threats.

In terms of protecting the individual, Martin Luther led the way. We noted earlier that in his effort to get princes and popes and emperors off the backs (or souls) of their subjects, Luther would exult, "Thought is free."[37] For the great reformer, faith is "secret, spiritual [and] hidden."[38] Individuals, in their capacity as souls trying to love God, cannot have their liberty of conscience curtailed by any worldly power. This sovereignty endows individuals with a certain dignity and inviolability. A person's thoughts about God are sacrosanct; they may not be subject to external trespass or coercion.

Having perused his Luther, John Locke would take another step toward the defense of the individual. He made the radical proposition that a church was "a *voluntary* Society of Men, joining themselves together of their own accord."[39] It's a stunning claim for that age – we come to God of our own free will, voluntarily. Faith cannot be forced upon us. The individual is at liberty to believe whatever she pleases. Likewise, she is at liberty to congregate with whom she wishes.

Locke and Luther erected a shelter for freedom of conscience. Accordingly, both defended the rights of dissenters and heretics against the power of the governing authorities. In so doing, they endowed the individual – even individuals espousing unorthodox beliefs – with protections against external coercion.

Thomas Jefferson and James Madison knew this tradition well. Jefferson reserved some of his loftiest rhetoric for this principle, such as when he exclaimed, "The rights of conscience we never submitted, we could not submit. We are answerable for them to our God."[40] Elsewhere he observed that "Almighty God has created the mind free."[41]

Madison seconded the idea when he averred that "the religion, then, of every man must be left to the conviction and conscience of every man; and it is the right of every man to exercise it as these may dictate."[42] People get to God as individuals, the (Protestant) architects say. Not satisfied to merely exalt the individual's rights of conscience, they were concerned with protecting the dissenting individual from the majority. Throughout his life Madison expressed the fear that the "majority may trespass on the rights of the minority."[43]

The architects feared the damage majorities could wreak not only on individuals, but on governing structures as well. As Philip Hamburger notes, Madison was deeply concerned that "America's churches would overpower the civil government." Interestingly, Jean-Jacques Rousseau expressed a nearly identical fear of what large groups (what he called "sectional associations") would do to the state. "Every citizen," Rousseau insisted, "should make up his own mind for himself," a sentiment shared by Madison.[44]

Madison and Jefferson's firewalls against the power of religious majorities included a hope that a plethora of religious groups, a "multiplicity of sects," would check and balance one another. In their conception, religious diversity would relieve pressure on the state. It was not for nothing that Jefferson exclaimed that "difference of opinion is advantageous in religion."[45]

And yet, ironically, today's evangelical movement *is* a multiplicity of sects. It is comprised of many Protestant denominations whose tens of millions share basic theological convictions *as well as* a loathing of the secular state. Evangelicals skillfully build coalitions of political convenience with Revivalist Catholics, Mormons, and Orthodox Jews (Francis Schaeffer used to refer to this as "co-belligerency").[46]

To respond to this challenge, a new generation of secular thinkers must reevaluate the multiplicity-of-sects approach. Diversity, in and of itself, is no longer an ally of the secular state, especially when the diversity coalesces in opposition to the fundamental secular principles of the state.

The architects' fear of religious communalism and majorities expanded to a much bolder and equally debilitating proposition in nineteenth-century liberal thought. Namely, that the individual is *morally and intellectually superior to the collective.* In the writings of John Stuart Mill (1806–1873), this impulse achieves perhaps its greatest expression. His classic essay "On Liberty" explores the dangers of stretching "unduly the powers of society over the individual."[47] He rails against "'the deep slumber of a decided opinion" and those who engage in "ape-like" imitation.[48] Among Mill's targets are "collective mediocrity" and "the general average of mankind" that is "moderate in intellect."[49]

Mill has a preference for Socratic heroes, the exceptional folks "abounding in individual greatness."[50] These iconoclasts are tormented by the very societies whose descendants will revere them.[51] Their martyrdom drives future creativity and innovation as it stretches the taut boundaries of human thought. Ergo, the just society has a compelling interest in protecting such individuals from the ignorant masses.

At its core, secularism retains an inveterate mistrust and fear of what masses and majorities can do. Individuals, by contrast, are viewed as both crucial and vulnerable. This is a powerful, even venerable, intellectual conceit. But it keeps running up against one basic political verity: in democratic societies, majorities, not individuals, retain the right to rule. Secularism's holy scriptures imbue the individual with so much sanctity that group formation seems like an afterthought, if not something of a mortal sin. But group formation is a sterling defense for secularism against groups that wish to vote secularism out of existence!

There is only so much that individuals can do in a liberal democracy. Socrates, after all, was sentenced to death by his countrymen. Also, people like Socrates aren't easy to mobilize. It is awfully difficult to find, let alone band together, fifty thousand Socrateses willing to march on the Washington Mall in the name of one issue. A robust celebration of the individual may often have a troubling corollary: the inability of those individuals to establish a collective capable of fending off groups with less admiration for individuals and their rights.

Revivalists, needless to say, share none of these misgivings about being a collective. There is no doctrinal hesitation or braking mechanism on their ability to engage in large-scale group action. Very little in their sacred texts urges them to nonconformity and individualism. They are rarely enjoined to express heretical thoughts.

Judaism and Islam also celebrate the idea of a community obedient before and under God. This impulse is expressed in the Islamic notion of the *umma* (meaning the community under God). Similarly, it is evident in the Jewish dictum that the messiah will come when every single Jew on earth observes the Sabbath twice consecutively; the importance of communal solidarity runs deep in many forms of religious thought.[52]

So we ask these questions: In a contest between the Revivalist espousing communalism and the secular believer or nonbeliever praising the exceptional individual, who gets to dance on election night? Who ends up in a body bag? Deep in the marrow of secularism (and liberalism) lie ideas about individuals and collectives that may be detrimental to the long-term health of the movement.

Secularism's Grand Strategy

Why did separationist secularism in the United States, fresh off of its dazzling midcentury judicial triumphs, buckle within a few decades of the Christian Right's emergence in the late 1970s?

Our analysis has identified a variety of factors, some related to secularism in general and others particular to its American variant. As regards the general factors, secular states *always* have a hard time fending off communalist charges. American secularism's days were probably numbered from the moment the Revivalists returned to the fray in the 1970s. Previous Protestant Revivalist movements in the United States had either supported separationism (because of anti-Catholic or Luther-like convictions) or been apathetic to it altogether.

By the final decades of the twentieth century, however, separationism was The Adversary. In truth, it never had much of a chance.

A high priority on secularism's to-do list should be gathering its intellectuals together for the purposes of grand strategy. Their assignment: to figure out how to combat the problem of lawful Revivalists who use democratic means to render our society less secular and eventually, by extension, less democratic. The answer they come up with can't involve using the courts to impose their agenda. The judicial strategy is no longer viable.

In terms of the American separationist movement that carried secularism's flag, we have identified a fairly obvious flaw. Truth be told, it wasn't much of a movement to begin with. The Christian Right has at its disposal tens of millions of white evangelicals and fundamentalists who call themselves Christians. Secularism, by contrast, does not (and never did) have tens of millions of people who called themselves secularists.

Instead, it had "adjective" members: secular Jews, secular teachers, secular Baptists, secular civil libertarians, secular pornographers, and so forth. Very few of them would describe their core identity as secularist. That was *not* a problem, actually. The stickler was that they did not think of themselves as secular in any way, shape, or form, despite behavior proving they valued secular ideals. For after all, outside of an obsession with the wall of separation, what values has American secularism promoted? Secularism does indeed stand for additional values. To secure its future, it will need to articulate them and inspire Americans to take up the adjective as part of their identity – to think of themselves as secular or secularish.

Successful political movements have intellectuals who lead the masses, and it is from these masses that legions of activists and operatives are drawn. The Revivalists had intellectuals like Schaeffer, operatives like Reed, and a seemingly bottomless supply of true believers ready to overrun PTA boards, work the phone banks, and canvas neighborhoods for fellow travelers. Secularism, undeniably, had its fair share of quality intellectuals, from jurists like Leo Pfeffer to the strategists working for Americans United. But it had no masses at its disposal. As for activists – the intellectuals themselves had to do double duty in this role! In truth, the judicial strategy was the only plausible tactic available to top-heavy, numbers-starved secularists.

Successful political movements have clearly articulated beliefs that bind together their diverse members. Evangelicals don't agree on everything. Yet there are core convictions – about the infallibility and inerrancy of the Bible, about accepting Jesus Christ as one's personal savior, about heterosexuality as normative and God-ordained – that unite its many denominations.

Secularism, by contrast, did not have an organic base of like-minded supporters. The members of its advocacy coalition have always had "little in common, ideologically" outside of a narrow commitment to separationism.[53] Secularism achieved some monumental gains in its heyday. But it lacked the leadership and the vision (and membership) to consolidate and eventually defend those gains.

This is something the Revivalists have exploited brilliantly. Their rants in the Reagan era depicted "secular humanism" as a frightful menace. Fundamentalists and evangelicals made a fundraising mint by depicting it as a national security threat. As a dean of the College of Arts and Sciences at Bob Jones University phrased it in 1987, "Humanism is a clear and present danger to every Bible-believing Christian. It is a satanic onslaught that must be fought in the power of the Spirit."[54]

Secularism shared disgrace with the Soviet Union by scoring high on the Antichrist meter. But Communism, at least, had a massive state apparatus at its disposal and thick and devious methods of coercion. But where exactly was the empire of American secularism? Where was its army? its impassioned hordes? its war chest?

There was none of that. Though there were elite units, so to speak. Secularism's defenders were stationed in print and television newsrooms. They worked in cinema and the domains of high art. Down on campus you had entire platoons of bespectacled secularists who sympathized with the cause. But those elites were either too unfocused, or too poorly led, or too weakly aligned with the movement, or too few, to repel the onslaught of the Revivalists.

There was, quite frankly, always a Wizard of Oz quality to the golden age of secularism in the middle of the twentieth century.

When Christian conservatives first breached and then penetrated the corridors of power, their post-conquest inventory may have surprised them. When they peeked behind the curtain, they realized that the secular leviathan they had always feared was something of a mythical beast.

8

Are Democrats Secularists?

> If your organization puts medicine in people's hands, food in people's mouths, or a roof over people's heads, then you're succeeding. And for the sake of our country, the Government ought to support your work.
>
> – George W. Bush, *"Remarks at the Office of Faith-Based and Community Initiatives National Conference," June 26, 2008*

On July 1, 2008, at a campaign stop in Zanesville, Ohio, Senator Barack Obama made a surprising announcement. The subject of his speech was George W. Bush's much-maligned Office of Faith-Based and Community Initiatives. Obama warmed up the crowd by discussing his "personal commitment to Christ" and reminiscing about his days as a community organizer back in Chicago. Then he took up his theme. The presumptive Democratic nominee for the presidency promised that if elected, he would not only retain the office but make it "a critical part of my administration."[1]

Some observers were stunned. After all, the Office of Faith-Based and Community Initiatives was a hallmark of the Republican policy product line. It reeked of the GOP's pandering to the Christian Right. The very mention of its name sent liberals on house-hunting expeditions to Manitoba. Worse yet, Bush's office was perceived in Democratic circles, but not only in Democratic circles, as a total mess. Why hadn't Obama simply told his Ohio audience that, if elected, he would executive-order the damn thing right out of existence?

The Democrats were the party of secularism, right?[2] Didn't a politician like Mario Cuomo represent their worldview? He was the Catholic governor of New York who took pains to distinguish his private dislike of abortion from his public responsibility to represent his constituents. He was the same governor who reminded us that he served many different Americas. In a speech delivered at Notre Dame in 1984, Cuomo declared, "I protect my right to be a Catholic by preserving your right to believe as a Jew, a Protestant, or non-believer, or as anything else you choose."[3]

But now in 2008 the Democrats were choosing something different; those who were familiar with the party's recent electoral woes understood that it was time to say goodbye to secularism (and whether secularism signified separationism or nonbelief or something else was not a distinction they were about to draw). After George W. Bush was reelected in 2004 the Democrats did a lot of "soul-searching."[4] Strategists conducting the autopsy of John Kerry's demoralizing defeat made two startling discoveries. The first finding was that—*who knew?*—there were a whole lot of religious people in the United States. The second was that the Democrats' alliance with secularism and separationism, which had lasted for nearly half a century, was yielding diminishing returns. In swing states brimming with those newly discovered religious people, secularism was no fan favorite.

Democratic pollsters also couldn't help but notice the role that the "values voters" played in Bush's 2004 reelection. These evangelicals, who constitute roughly one-quarter of the electorate, gave nearly 80 percent of their ballot to the incumbent. In fact, there is reason to suppose that the evangelical vote in Ohio tipped the state, and therefore the entire election, to the red side of the ledger.[5] The evangelicals' energy, enthusiasm, and election-day performance carried Bush through to a second term.

Energy, enthusiasm, election-day performance—these are not terms associated with secularists. After 2004, party strategists could no longer neglect the fact that the so-called secular base of their party was underperforming. Candidates who were strongly associated with separationist ideals had been a disappointment. The failed presidential runs of Walter Mondale (1984), Michael Dukakis (1988), and John Kerry (2004) had imparted a valuable lesson: there was a correlation between supporting candidates who were card-carrying secularists and watching those same candidates sit through the inaugural address of their Republican opponent. The Democrats were determined not to repeat that mistake.[6]

It thus made perfect sense that each of the major Democratic presidential hopefuls in 2008 was eager to engage in "faith and values" campaigning. John Edwards – who was apparently not right with God at the time – was everywhere citing Matthew 25 as a proof text for his antipoverty policies. Hillary Clinton relentlessly stressed her religious bona fides on the campaign trail.

Yet the former First Lady was, to her credit, one of the few Democrats to express something akin to a pang of secular conscience. At the April 2008 Compassion Forum – an event featuring her, Barack Obama, a gospel choir, dozens of faith leaders posing questions, CNN, and the campus of Messiah College – she reflected thoughtfully on the whole novel scenario: "I understand why some people, even religious people, even people of faith might say, why are you having this forum? And why are you exploring these issues from two people who are vying to be president of the United States? And I think that's a fair question to ask. I am here because I think it's also fair for us to have this conversation. But I'm very conscious of how thoughtfully we must proceed."[7]

Obama, of course, had no compunction about having this conversation. His political capital in the party was built on his unique ability to help the Democrats overcome their addiction to secularism. He was the Senate candidate who, back at the 2004 Democratic National Convention, memorably intoned, "We worship an awesome God in the Blue States."[8]

As he grew in stature, Senator Obama had the awesomely peculiar habit of using the religion card against his own party. In *The Audacity of Hope* he accused the Democrats of becoming the "party of reaction": "In reaction to religious overreach, we equate tolerance with secularism, and forfeit the moral language that would help infuse our policies with a larger meaning."[9] By 2008, then, both Obama and the Democrats had completely rethought their relation to the so-called secular base. Supersizing Bush's Office of Faith-Based and

Community Initiatives was a swell way to signal to the electorate, and of course the evangelicals, that he took religion seriously.

Barack Obama remained true to this word. Within days of his inauguration he renamed and renovated the old Bush bureau. So it came to pass that the White House Office of Faith-Based and Neighborhood Partnerships was brought into being on February 5, 2009, by an executive order of our forty-fourth president, a Democrat.[10]

When confronted by challenges like the existence of this office, secularists need to understand precisely what they signify. The Bush-Obama contraption does, undoubtedly, pose a challenge to separationist ways of thinking. It does not, however, necessarily suggest the onset of an American theocracy. Truth be told, the office can *plausibly be considered a form of secularism.*

If that sounds strange, it's because in America we don't really discuss secularism outside the separationist paradigm. All other alternatives, and there are many, are ignored. The consequence is that there is no middle ground in secular advocacy.

Those who champion the office appeal to a doctrine known as accommodationism.[11] It may – emphasis on *may* – be warranted to view this approach as a form of "soft secularism." A study of the office will permit us to tally up accommodationism's pros and cons. It will also position us to distinguish sincere versions of this approach from pious frauds.

Bush and Charitable Choice

The genesis of the office is an intriguing story in and of itself. It occurred somewhat by accident and on yet another Democrat's watch. It came into being when the Clinton administration let stand the "charitable choice" provision in a larger welfare-reform bill known as the Personal Responsibility and Work Opportunity Reconciliation Act of 1996.[12]

Little did Bill Clinton and Al Gore (who supported the provision in his 2000 run) know that charitable choice could do so much damage to the separationist status quo. It was like a refrigerated stem cell.

After lying dormant for a few years it would be harvested into an entirely new policy organism by Republican lab technicians.[13]

This legislation is complex, but at its core charitable choice "allows the government to fund church social services directly and explicitly protects the religious character of such organizations that accept public funds."[14] Charitable choice was, in the words of one scholar, "a bit of a 'sleeper.'"[15] It seemed like a very minor clause back in 1996. Few knew it was even on the books.[16]

One person, however, was acutely aware of its existence – a certain Texas governor by the name of George W. Bush. In his 1999 autobiography, *A Charge to Keep,* he lamented the fact that although Texas had "many kind and loving people who follow a religious imperative to help neighbors in need," the state made it difficult for them to do so.[17] In order to rectify that problem, Bush signed an executive order encouraging government agencies in the Lone Star State to smooth the path for religious groups that might be qualified to provide social services.

Teen Challenge of South Texas was precisely such a group. Bush was quite fond of this provider of drug and alcohol treatment. After all, the group was, in his words, "saving lives through the transforming power of faith."[18] State regulators in Texas may have been concerned that its course of therapy "focused on Bible study and prayer."[19] But not the born-again Governor Bush. With the charitable choice provision, he found the legal blocking tackle that permitted him to hoist taxpayer dollars into the coffers of faith-based providers.

Upon assuming the presidency in 2001, Bush was eager to get to work on his Office of Faith-Based and Community Initiatives. So eager, that the very first executive order of his administration addressed the matter. It reads as follows: "The paramount goal is compassionate results, and private and charitable community groups, including religious ones, should have the fullest opportunity permitted by law *to compete on a level playing field,* so long as they achieve valid public purposes, such as curbing crime, conquering addiction, strengthening families and neighborhoods, and overcoming poverty."[20]

The office always was and still is a very difficult thing to comprehend. Bush's version, for example, initially sprawled across five federal agencies (soon it became eleven).[21] Critics, advocacy groups, journalists, and lawmakers themselves have often wondered what exactly the office was doing.[22]

Questions have lingered for nearly a decade – across both the Bush and Obama administrations – about where its budget comes from, how much money it has received (apparently $2.1 billion in 2005, to give a sense of the scope of the thing), and where its budget actually goes.[23] Debates have raged as to how outcomes are assessed for faith-based service providers that offer prayer and scripture reading as their program of therapy. Transparency has never been the faith-based office's strong suit, whether it is run by a Republican or Democratic regime.[24]

Readers might spare themselves some vexation if they stopped trying to understand the incomprehensible mechanics of the office. It would be better to focus more on what it represents: something which American secularists had better think about long and hard. As one original Bush staffer proudly put it, the office aspired to enact "a determined and sweeping reframing of the relationship between government and civil society."[25]

The scholar Mary Segers confirms the notion that a monumental shift has taken place. In her estimation the faith-based office represents "a *strategic* change in White House thinking about the relationship between church and state – from a traditional one based on *separation* to a new one characterized by *collaboration*."[26]

Accommodationism in Theory

Secularism is a political philosophy about how a government should relate to religion so as to maximize order, freedom of religion, and freedom from religion. Separationism, *which is one variant of this philosophy,* maintains that all three outcomes will be achieved in spades if there is, in effect, *no* relation between government and religion.

The Bush-Obama faith-based offices are legitimated by another variant, a reading of the Constitution known as accommodationism.

To give a law textbook definition, accommodationists "argue that the government may aid religion as long as it does not do so to the advantage of one religion over another . . . aid must be given in a nonpreferential manner; what is available to one religion must be available to all."[27]

Accommodationist (sometimes called nonpreferentialist) approaches have come to the fore relatively recently.[28] They tend to hinge on highly controversial readings of the First Amendment and early American history. Separationists, as we have seen, refract the Establishment Clause through the prism of Jefferson's 1802 Danbury letter. Accommodationists take their own crack at constitutional ventriloquism. They tend to read into the silences of the clause. Their argument is built on what the Establishment Clause *doesn't* say a government *is not allowed* to do. If this seems like a spurious legal argument, it certainly is. But then again, what other recourse do jurists have when confronted by the word-and-thought-defying concision and ambiguity of the religion clauses?

After all, how can a nation of 300 million people and countless creeds regulate relations between its government and its many faiths on the basis of marching orders as sparse and imprecise as these: "Congress shall make no law respecting an establishment of religion, or prohibiting the free exercise thereof"?[29]

When confronted with uncertainty, it is always expedient to assign blame and in this case it may be assigned to the Framers of the Constitution. One legal scholar notes that the religion clauses raise "problems ranging from paradoxical to impenetrable" and laments the "somewhat hasty and absentminded manner" in which Congress approached the text.[30] Another speaks of its authors as "vague if not careless draftsmen" and wonders if they shared a "single understanding" of its meaning.[31] "Inexorably unclear in its intention" is how still another judges it.[32] A flustered analyst despairs that the disheveled Establishment Clause may have "no original intent at all."[33]

Any attempt to make sense of it all must begin with one trying term, *establishment*. What precisely did the Framers understand that to mean? A historical inquiry reveals that in early America an establishment often entailed a "legal union of government and religion."[34] The principle of establishment sounds fairly simple. Yet the actual arrangements on the ground that existed in colonial and post-revolutionary America were varied and complex, livable and unlivable. Thus, when the Framers thought of establishment, a whole range of connotations may have come to mind.

Some may have thought of *exclusive* establishments, in which one religion and one religion only was associated with the governing authority. These establishments often evoked bad memories. The Founders may have remembered, for example, Virginia's infamous "Dale's Laws" of 1610–1619. This legal code entailed public flogging for religious dissent, mandatory daily church attendance, and bodkins cast through the tongues of blasphemers. These punishments were enforced through an "ecclesiastical policing network" that inspired a number of settlers to run away or commit suicide.[35]

Then there were colonies (and later, states) that featured what are now called *multiple* establishments. Here we witness what one scholar calls "nonexclusive assistance to all churches."[36] For example, pre- and postcolonial Massachusetts featured arrangements whereby payment of taxes for the maintenance of a church or clerical salaries was mandatory. The citizen, however, could direct those funds to support the Christian denomination of his choice. In addition to the exclusive and multiple establishments just mentioned, some towns sported a *dual* establishment.[37]

In response to all of this, the separationist might offer this riposte: "Well, whatever *establishment* meant back then – exclusive, multiple, dual – the Framers didn't want any part of it!" True enough. Yet a crucial proviso is in order. While it is abundantly clear that the Founders intended for the federal government (that is, Congress) to stay clear of an establishment of religion, it also seems apparent that they thought differently about *states*.[38]

It is, in fact, likely that many had no opposition whatsoever to state establishments. Some scholars suggest that the main purpose of prohibiting federal establishments was to ensure the right of states to make their own decisions about the proper place of religion in public life. The feasibility of that interpretation might be gleaned from the curious fact that seven of fourteen states had establishments after the Constitution was ratified in 1791.[39] The last official establishment, in Massachusetts, persevered *until 1833*. In contrast to federal establishments, state establishments clearly had some modicum of popular support.

The Establishment Clause is a blunderbuss of a foundational document. In its brevity and ambiguity, it doesn't really tell us how government is supposed to engage with religion outside of not *making laws about* establishment (that itself is a peculiar phrase, when you think about it).

The rationale for the Bush-Obama faith-based office hinges, in part, on a fairly strained accommodationist reading of this problematic clause. This reading has gained momentum since the 1980s; its basic premises are encapsulated in Justice William Rehnquist's dissent in the 1985 *Wallace v. Jaffree* case.[40] His intervention signals one of the most forceful challenges issued to the separationist approach and perhaps the beginning of its drawn-out judicial demise.

Mincing no words, Rehnquist argued, "The Establishment Clause did not require government neutrality between religion and irreligion nor did it prohibit the Federal Government from providing nondiscriminatory aid to religion. There is simply no historical foundation for the proposition that the Framers intended to build the 'wall of separation' that was constitutionalized in *Everson*."[41]

Accommodationism, then, interprets the Establishment Clause to say that a government *can* support religion, materially or in some other way. A government can even collaborate with religion. It can do so on the condition that it not show preference to any *one* religion. That's why accommodationists can claim that they are disestablishmentarians in the truest sense of the term. They oppose an exclusive federal establishment, just as the Founders did.

Notice how this interpretation "reads" the silence. Justice Rehnquist concluded that the religion clauses don't prohibit "nondiscriminatory aid" to religious institutions. But the religion clauses don't prohibit a lot of things, and that's because the religion clauses are exactly sixteen words long.

The accommodationist approach is reminiscent of the multiple-establishment approach. Since a few of the Framers had no difficulty with government support for all (Protestant) religious denominations, some think a similar policy would pass constitutional muster today if it was broadened to include more religious groups. It's a surmise somewhat hampered by the fact that the First Amendment says no such thing.[42] Nevertheless, this reading is undergirded by the old civic republican premise (see Chapter 2) that religion *in general* provides valuable advantages to the social body.[43] Why accommodate religion? Because positive benefits will accrue to all.

For separationists, governmental evenhandedness is achieved by maintaining a (great) distance from all religions. Accommodationism inverts that logic: the government maintains a benevolent proximity to *all* faith-based groups.

So, *in theory,* accommodationism could be seen as a soft form of secularism. Now let's get back to the office and see how one attempt to put this theory into play worked out.

The Critique

As a person who overcame alcohol dependency, George W. Bush personally witnessed faith's ability to alleviate human misery. As president he wanted other Americans to experience the wonder-working power of religion. By promoting faith-based initiatives, one of the "signature ideas" of his campaign, he believed he had found the proper delivery mechanism.[44]

Part of Bush's "compassionate conservatism" consisted of letting religious groups deliver their therapeutic services to those in need. Yet it bears noting that the federal government had been funding religious groups that deliver these services *for decades.* Respected organizations such as Catholic Charities, United Jewish Communities, and Lutheran Social Services, among others, provided invaluable aid to everyone from homeless people, to prisoners, to those addicted to drugs and alcohol.[45]

These "religiously affiliated" groups, as they are called, played by the rules of the old separationist paradigm. They made sure to separate, or wall off, their social-service mission from their religious mission. The agencies were "essentially nonreligious, having nonreligious governing boards, running on carefully segregated financial accounts, and operating in physical environments where the religious trappings and imagery had been removed."[46]

In short, these charities had *secularized themselves* in order to qualify for federal monies. They developed discrete social-service arms so as not to trespass on church-state boundaries. In an age of separationism, there were many things they were not allowed to do. Catholics couldn't approach an alcoholic in one of their programs and ask her to pray. Lutherans couldn't request that drug addicts in their care peruse the fifty-five volumes of Luther's writings. Jews running a soup kitchen with the aid of taxpayer dollars couldn't subject their clients to lectures about Rabbi Akiba's teachings. None of the providers could adorn their offices with ostentatious religious symbols. Nor could they hire staff members solely on the basis of their religious views. No major hardship there.

This is because the biggest players in this field were "of the modernist stripe."[47] In other words, the types of religious people who performed these good works were moderates and secularish folks whose theology was "ecumenical and humanitarian."[48] They operated in harmony with the general separationist civic consensus that prevailed in America in the second half of the twentieth century.

Needless to say, many of the groups the Bush administration sought to empower (and fund) were not "of the modernist stripe." In fact, many were Christian Revivalists. Legal scholars use the term "pervasively sectarian" to describe these believers. Pervasively sectarian groups cannot and *will not* wall off a social-service mission from its religious component.[49] In other words, their religious convictions and practices are part of the service they offer. These groups insist that their deliverables are effective for precisely this reason. As one commentator put it, "Sharing of religious convictions *by* openly religious people *in* openly religious settings was at the core of effective treatment and rehabilitation."[50]

Let's set this in the context of accommodationist theory. Prior to Bush, the federal government would accommodate religious social-service providers who behaved in a secular manner (that is, groups that did not connect their good works with a religious message, missionizing, and so forth). The government said, "We will support your do-goodery only if you cast aside your particular religious scruples as you do good." In the new version, the government says, "We will support your do-goodery no matter what your religious scruples are, even if those scruples may be illegal and discriminate against entire categories of American citizens. Now, go out there and do good!"[51]

Illegal? Discriminatory? Here we come to the greatest difficulty confronting the office. Under the 1964 Title VII exemptions, religious groups are free to hire only those who share their religious convictions.[52] That means that in their capacity as religious groups they are free to discriminate (with certain restrictions) in their choice of employees. The problem begins when we consider that in the view of many Revivalist groups, *all* gay people do not share their religious convictions. In accord with Title VII, a gay person could be refused employment at a soup kitchen operated by a fundamentalist church.

But what happens when the government is funding that very soup kitchen through the intermediary of a faith-based group? It is one thing to permit a religious group to discriminate in its hiring practices out of respect for religious liberty. It is another thing entirely for the government to bankroll and hence indirectly endorse that discrimination. To invert Justice Rehnquist's words, this is quite literally *discriminatory aid.*

In July 2001 it was reported that the Salvation Army called the White House to request an exemption from obeying laws prohibiting sexual-preference discrimination in job hiring.[53] Once stories of this nature made the rounds in Washington, critics pounced. One Democrat warned that the fundamentalist Bob Jones University would "be able to take federal dollars for an alcohol treatment program and put out a sign that says no Catholics or Jews need apply here for a federally funded job."[54] The Bush administration ultimately turned down the Salvation Army's request.[55]

In 2001, legislative approval for Bush's faith-based initiatives stalled in the Senate and the discriminatory hiring fracas is probably why.[56] In return, Bush started bypassing Congress, implementing his own "executive strategy." After his setbacks in the legislative branch, the president quietly imposed his agenda via executive order.

Obama, incidentally, was quite cognizant of the uproar over the hiring issue. In his Zanesville address, he made sure to point out that Bush's operation had not "fulfilled its promise."[57] Were he elected, those who received funds from his faith-based office would not be permitted to discriminate in hiring: "If you get a federal grant, you can't use that grant money to proselytize to the people you help and you can't discriminate against them – or against the people you hire – on the basis of their religion."[58] Yet nearly throughout his entire term this promise has remained unfulfilled.[59]

As for Bush, the concerns over employment law were just the start of his problems. From day one of the office's existence (which was the ninth day of Bush's presidency!) the unit became an inviting target for progressives and liberals, who viewed it as yet another pilgrim step toward the advent of an American theocracy. These critics couldn't have been more gleeful when the office experienced a string of early setbacks.

There was the resignation, after just seven months on the job, of the office's mercurial director, the frank and forthright John DiIulio.[60] Later this iconoclastic professor from the University of Pennsylvania made a point of identifying the office as a political patronage machine, referring to it as the "reign of the Mayberry Machiavellis," where everything "was run by the political arm."[61]

The next director was far more on-message. DiIulio's successor, Jim Towey, railed against "secular extremists."[62] He mocked pagans when he was queried as to whether they too would be able to receive federal funds to help the poor.[63] Breaking the cardinal rule of civic republicanism – "Thou shalt not disparage any religion" – Towey proceeded to tar them as a "fringe group" and imply that they lacked "loving hearts."[64]

But the most damning witness was the former Bush staffer David Kuo. In his riveting 2006 tell-all, *Tempting Faith: An Inside Story of Political Seduction,* the number two in Bush's faith-based office told a tale of a unit that was understaffed, overpoliticized, and willfully duplicitous. Making his testimony particularly compelling was the fact that Kuo was himself an evangelical Christian and admirer of President Bush.

Tempting Faith regales us with astonishing White House human-interest stories. Particularly memorable is a recollection of Karl Rove chasing down a member of the transition team and excoriating him with the words "Just get me a F*%#ing faith-based thing. Got it?"[65] Equally newsworthy was the accusation that many White House staffers thought of their frequent evangelical visitors as "nuts," "boorish," "ridiculous," "out of control," and "goofy."[66]

Kuo lent credence to a widely held complaint: namely, that the office was nothing more than a patronage shop for rewarding constituencies loyal to Bush. It has also been alleged that Bush's office was designed not just to level the playing field, but to level the Democratic Party – its real objective was to crystallize Rove's strategy for reelecting Bush in 2004. Critics charged that the actual purpose of the office was to lure white evangelicals into the Bush fold.

Another objective of the faith-based office may have consisted of cutting into the Democrats' traditional stranglehold of African American voters. By reaching out to the black church, the GOP hoped to chip away at one of their opponents' most loyal constituencies.[67] Evidence from the 2004 election, incidentally, in the decisive state of Ohio suggests that it may have accomplished its goal.[68]

Kuo and his colleagues staged countless meetings with clerics and other religious leaders. The manifest objective of these get-togethers was to teach such groups how to access the alleged billions of dollars that the government had placed on the table. Critics, for their part, couldn't help but notice that most of these gatherings took place in the swing states Bush would need for reelection in 2004.[69]

The entire operation, Kuo laments, morphed into a "sad charade, to provide political cover to a White House that needed compassion and religion as political tools."[70] He ends his book with a plea for evangelical Christians to begin a "fast" from politics and focus on their souls.[71]

The Obama office has been less roiled by such controversies. One of the great accomplishments of its director, Joshua DuBois, has been his uncanny ability to stay out of the headlines – so much so that the office appears to be flying almost completely under the radar. Whispers surfaced that Obama's office was being used in his effort to enact health-care legislation. But no national outcry on the order of those heard in the Bush era has afflicted the camera-shy Obama office.[72]

The Obama administration, however, has shown signs of being inclined to some modicum of fair play. It did, after all, appoint the Reverend Barry Lynn of Americans United for Separation of Church and

State to its advisory board for one term. Lynn is one of the most committed and knowledgeable separationist thinkers in America. That DuBois was willing to invite him in is a sign that he is willing to listen to his critics. It also serves as a model. In order to understand what is happening in Washington, secularists need to work with the power, rather than fight it.

Can Secularists Live with Accommodation?

Upon launching its new iteration of the faith-based office in 2009, the Obama White House proclaimed, "As the priorities of this Office are carried out, it will be done in a way that upholds the Constitution – by ensuring that both existing programs and new proposals are consistent with American laws and values. The separation of church and state is a principle President Obama supports firmly – not only because it protects our democracy, but also because it protects the plurality of America's religious and civic life."[73]

Let's be clear: for the Obama administration to say that the office upholds the separation of church and state is utterly preposterous. The office marks another milestone in a continuing, decades-long shift away from separationist secularism to the accommodationist paradigm. The latter could be considered secular in that it permits no establishment of religion. On the basis of what we have seen in the office's first decade, however, there is reason to be skeptical about this softer secularism.

Secularism, as we define it, must ensure freedom *from* religion. By that standard, accommodationism has some in-built deficiencies. Any government beholden to the conviction that religion is an unambiguous force for good must inevitably have a far less positive assessment of those who have no religion or reject religion. Accommodationism doesn't even possess the conceptual vocabulary to deal with atheists, agnostics, freethinkers, and so forth.

As we saw, Justice Rehnquist mocked the idea that the Establishment Clause required the government to be neutral as concerns both religion and the absence of it. The office and its supporting accommodationist theory are so giddy in their always-sunshine evaluation of religion that it draws a tacit distinction. Believers are essential. Nonbelievers: meh.

This is a fairly straightforward problem to solve (though given the pervasive dislike for atheists in America, it may take some doing). A president could pen an executive order to the effect that a faith-based office will work with organizations regardless of their faith or lack of faith. It has yet to happen and it probably won't any time soon. But if President Obama could acknowledge in his inaugural address that America is also a nation of nonbelievers, he could surely deputize a group such as American Atheists to run a homeless shelter on the federal dime.

A more complex dilemma for accommodationism concerns the idea of "excessive entanglement." This phrase first emerged in *Walz v. Tax Commission* in 1970.[74] It resurfaced as the third prong in the infamous *Lemon v. Kurtzman* case of 1971.[75] In that decision, Justice Burger reasoned that a law could not foster "an excessive government entanglement with religion."[76]

The entanglement test has lost favor of late. Still, how can the faith-based office *not* lead to excessive entanglement? With hundreds of millions of taxpayer dollars being sluiced into church coffers, at some point a Mr. or Ms. Fiddle over in federal accounting is going to have to ask some questions: Is the recipient using the money honestly? Is it obeying federal laws? Is it providing the services it promised to provide? Are these services effective? The list of concerns could be lengthy.

Soft secularisms create massive entanglement dilemmas. They lack the theoretical simplicity of separationist views that sever church from state. Once the latter develops a fiduciary relationship with the former, it must engage in invasive monitoring behaviors (such as checking the books, observing hiring protocols) unprecedented in American religious history.

A related drawback of soft secularism concerns odious religious groups that the government might not really want to draw into partnership. What is to be done with white supremacist religions that seek federal funds to offer counseling to families who have defaulted on their mortgages in Idaho? Insofar as the government works on a nonpreferential basis, how can it possibly draw distinctions between “good” and “bad” religious groups? What about a Christian clinic trying to cure gays of their “affliction”? Or a Christian pregnancy counseling center that receives federal monies to dissuade clients from their federal right to a legal abortion?

On the basis of these concerns, accommodationism presents secularists with a lot to worry about. Yet once again the separationist is on the outside, looking in. Programs like the office, surveys indicate, are actually quite popular. In 2001 a Pew Research Center study discovered that 75 percent of Americans favor faith-based initiatives (69 percent when the study was replicated in August 2009).[77] Here is a problem we have encountered again and again: separationists are not down with the people.

So what is to be done? Secularists are more than welcome to dig in and fight all three branches of the government, two political parties, and the American public as well. They can even be equipped with a suit of armor and a lance if they think it will be helpful. Yet pragmatism recommends a different course of action. In this book’s conclusion we will look at some ways in which a revived secularism might use accommodationism to its advantage.

Then again, accommodationism may be the least of secularism’s worries. The GOP has become enthralled by lethal *pseudo*-accommodationist approaches. These, as we are about to see, are shills for a growing Christian nation ideology.

9

The Christian Nation and the GOP

> Christianity is the religion that shaped America and made her what she is today. In fact, historically speaking, it can be irrefutably demonstrated that Biblical Christianity in America produced many of the cherished traditions still enjoyed today . . .
>
> – David Barton, *WallBuilders website*

One might imagine that an organization named WallBuilders would be a separationist advocacy outfit. Or maybe an artists' collective honoring the memory of Justice Hugo Black through public performances and group shows. Nothing could be further from the truth. Founded in 1989 by the evangelical Christian David Barton, WallBuilders describes its mission as follows: "(1) educating the nation concerning the Godly foundation of our country; (2) providing information to federal, state, and local officials as they develop public policies which reflect Biblical values; and (3) encouraging Christians to be involved in the civic arena."[1]

The indefatigable Barton is the most prominent, but certainly not the only, advocate of a worldview that is gaining increasing traction among Christian Revivalists. In his elucidating book *Was America Founded as a Christian Nation?* the historian John Fea discusses the rise of evangelicals who contend that "the study of the past . . . has been held hostage by secularists who have rejected the notion that the American founders sought to forge a country that was Christian."[2]

By now readers will not be surprised to learn that where there is conservative Christian smoke there is activist fire. The spry Revivalists are all over the Christian nation idea.[3] Some insist that America was actually established as a Christian nation. Others are taking concrete steps to restore a fallen America to that cherished status. Barton, for his part, claims that the Founders did not want to remove Christianity from public life but merely intended to stop "the establishment of a single national denomination."[4]

This large and diverse activist pool has national leaders, circuit lecturers, Web-savvy outreach, fundraising arms, literature to peddle, and DVDs too. Most important, these Revivalists have intimate links to the GOP. Barton himself served as vice chairman of the Texas Republican Party from 1997 to 2006.[5] During that period he did a tour of duty mobilizing Christian voters for George W. Bush's 2004 presidential run.[6]

The name WallBuilders is taken from passages in the biblical book of Nehemiah that chronicle the reconstruction of Jerusalem's dilapidated ramparts.[7] The group's website proclaims, "We have chosen this historical concept of 'rebuilding the walls' to represent allegorically the call for citizen involvement in rebuilding our nation's foundations."[8]

Much WallBuilder rhetoric centers on making the *historical* case for the Christian founding of this country. This historical analysis drives the group's policy prescriptions. Former Arkansas governor Mike Huckabee has referred to Barton as "maybe the greatest living historian on the spiritual nature of America's early days."[9] The liberal organization People for the American Way, apparently not deferring to Mr. Huckabee's scholarly assessment, had a different opinion. They described the uncredentialed Barton as "the Right's favorite pseudo-historian."[10]

Barton and those like him draw much of their energy and inspiration from a historical conundrum we have been charting throughout this book. For despite our great nation's insistence on prohibiting an establishment of religion, *it has often acted as if it were a Christian (meaning Protestant) nation beholden to the moral and political principles of the Protestant Christians who govern it.* This is an absolute doozy of a contradiction. It has fueled intense scrums between Revivalists and separationists about the proper place of religion in public life.

If the debate seems endless and insoluble, it is because each side has convincing proof texts. Secularists can point to the "godless Constitution," the Establishment Clause, and the Danbury letter.[11] To bolster their case they might invoke Article 11 of the Treaty of Tripoli (ratified in 1797). That document stipulates that "the government of the United States of America is not in any sense founded on the Christian Religion."[12]

The Revivalists, for their part, have scads of resources for pushback. Truth be told, they may have a numerical advantage here. In response to the secularists' deployment of the Treaty of Tripoli, they might raise the stakes with the Northwest Ordinance of 1787. That's a text whose third section begins with this lofty locution: "Religion, morality, and knowledge being necessary to good government . . ."[13] As for the godless Constitution, Revivalists theatrically invite secularists to show them where it mentions separation of church and state. They will interrupt the subsequent prolonged silence by invoking the Declaration of Independence, with its references to "Nature's God," "their Creator," "the Supreme Judge of the world," and "Divine Providence."[14]

On a roll, Revivalists can then crack open their treasure trove of quotes by the Founders of the republic.[15] The roster is impressive and long. George Washington – whose personal religious beliefs have been debated for more than two centuries – exclaimed in his Farewell Address, "Of all the dispositions and habits which lead to political prosperity, religion and morality are indispensable supports."[16] John Jay spoke of the "privilege and interest of our Christian Nation to select and prefer Christians for their rulers."[17] "A decent respect for Christianity," pronounced John Adams in his inaugural address, is "among the best recommendations for the public service."[18]

A sober and principled historian like Fea, incidentally, could agree that "between 1789 and 1865 Americans – North and South, Union and Confederate – understood themselves to be citizens of a Christian Nation."[19] A Christian himself, Fea makes the nuanced point that a country with such a dismal record on race relations never actually lived up to the challenge of being a Christian nation.

Recent research, then, has identified a staggering paradox. On the one hand, the American Founders did not want a national religion. On the other, they often indicated that Christianity *was* our national religion and that Christian beliefs were a prerequisite for holding elected office. The constitutional scholar Thomas Curry has gotten to the gist of the matter by noticing the complete disconnect between the theory and the practice of the Framers. In theory, there was to be no federal establishment. In practice, however, countless men of that generation "assumed that theirs was a Christian, i.e., Protestant country," which would "uphold the commonly agreed on Protestant ethos and morality."[20]

Christian nation activists are well aware of recent historical studies that have brought this paradox to the fore. Secularists, by contrast, have been slow to absorb this research or work through its practical implications. This is unfortunate because the practical implications are significant for secular activism. Decades back, rampaging conservative Christians were stopped dead in their tracks by the contention that ours was a nation founded on the principle of separation of church and state. But nowadays this silver bullet no longer conclusively neutralizes Revivalist claims.

Separationism as both a historical argument and a policy prescription has lost ground. We have seen that this approach lacks constitutional sanction, lengthy historical precedent, sympathetic jurists, friends in high places, and throngs of supporters. With separationism under siege, the question of religion's proper relation to government has been blown wide open. Secularists and Revivalists are competing to convince three branches of government, two political parties, and the American people as to how church and state should interact. Revivalists apparently are winning.

There is no political antidote to the bewildering swarm of Christian nation thrusts and parries that are menacing American secularism almost daily. Until reinforcements arrive in the form of more people and fresh ideas, today's secularists will just have to "hold the fort." One way of fending off the Revivalist assault consists of doubling down on old separationist arguments. But this strategy appears doomed. The time has arrived for pragmatism, for abandoning old positions, for stepping back from the wall of separation.

Playing the Secular Card: Laying Back

Christian nation ideology is, for some, a sincerely held conviction. For others, it is a form of demagoguery. John McCain's sudden and ill-advised "Christing up" during his 2008 presidential run is a neon-lit example of the latter.

In the fall of 2007, with his campaign in the doldrums, Senator John McCain of Arizona turned a few heads when he mentioned to a reporter, "The Constitution established the United States of America as a 'Christian Nation.' But I say that in the broadest sense."[21] The salty old sailor wasn't exactly known as the type of fellow who offered rhetorical sops to evangelicals and fundamentalists, even in the broadest sense. Yet now he was reading the First Amendment's Establishment Clause as if he were an ultraconservative Christian Revivalist. There was, as you may imagine, some backstory here.

McCain had, in fact, been feuding with the Christian Right for years.[22] His most notable battle (and defeat) took place seven years earlier, in the 2000 Republican presidential primary in South Carolina. The skullduggery that occurred there is still not fully understood. As best we can tell, some local operatives conspired to defame the senator from Arizona by spreading rumors that he had fathered an illegitimate interracial child.[23]

Aside from handing a crucial primary victory to George W. Bush, these machinations drove the notoriously hotheaded war hero over the edge. In the aftermath of his Palmetto State drubbing, McCain delivered the speech that struck the deathblow to his once promising campaign. "Neither party should be defined," thundered the senator, "by pandering to the outer reaches of American politics and the agents of intolerance, whether they be Louis Farrakhan or Al Sharpton on the left, or Pat Robertson or Jerry Falwell on the right."[24]

Robertson and Falwell on the Right must have been as astounded as they were incensed by all this "straight talk." Upbraiding the leaders of conservative Christendom was a feat that few Republicans in the post-Reagan era ever dared to perform. Fewer still drew scathing caricatures of them as McCain did in his 2003 book *Worth the Fight-*

ing For. In that same book the senator referred to himself as a "secular politician."[25]

But by the time 2007 rolled around, presidential hopefuls wanted to be called "secular politicians" about as much as they wanted to be called "martyrs for glorious jihad." McCain's strategists had learned the lesson of 2004: evangelicals, not secularists, held the keys to the White House. So in 2007 a few winged words from a desperate McCain about a Christian establishment made perfect sense. And if those words induced aneurisms in the secularists, that was collateral damage the campaign could live with.

Just to hammer the point home, the senator went on to express doubts about whether he would be comfortable with a Muslim as president.[26] Putting in his bid to win the Triple Crown of objectionable faith-based pandering, he proceeded to convert from the Episcopal Church to the voter-rich Baptist faith. (And he did so at the altar of a press conference, no less!)[27]

Enter, as if on cue, those defending secular ideals via op-ed columns. They now thrust "the maverick" into the limelight.[28] They excoriated the senator and his unconstitutional posturing. Their chief talking point: far from being a Christian nation, the United States was founded on the principle of separation of church and state.[29]

A *New York Times* op-ed writer cautioned McCain that "believers are to be wary of all mortal powers."[30] The *Kansas City Star* astutely remarked that "'Christian nation' . . . is insider jargon for people who consciously or unconsciously support a theocracy in which they get to be God's interpreters."[31] Americans United for Separation of Church and State reminded McCain that "the U.S. Constitution is secular and it creates a secular government that protects the rights of all citizens, religious or not."[32]

These well-meaning secularist interventions had an unintended consequence: they drew attention to the fact that McCain actually shared a good deal of common ground with evangelicals and fundamentalists. After all, he possessed impeccable pro-life credentials. He could point to a far better track record on the abortion issue than two of the candidates he trailed at the time: the master of flip-floppery Mitt Romney and the pro-choice Manhattanite Rudolph Giuliani.[33] The Christian Right lowered its guard.

Through cunning misdirection, shamelessly divisive rhetoric, and the predictable overreaction of the separationists, McCain was able to revive his moribund candidacy. By the time of the nominating convention, the maverick had done a respectable job of placating the conservative Christian column of the GOP base that had been highly suspicious of him. With his eventual selection of a running mate – Sarah Palin, a more credible and on-message Christian nation advocate – he finally energized that base.[34]

McCain's presidential run is a thing of the past, but the tensions it exposed remain with us. Appeals to the ideal of a Christian nation, as McCain's handlers surely knew, had a receptive audience. A poll conducted in 2007 found that 55 percent of Americans thought that the Constitution established the United States as just such a nation.[35] A whopping 65 percent thought that the Founders intended it to be so.[36]

One possible lesson to be drawn from the McCain episode is that separationist intervention has negative effects. The position tends to alienate voters. It is difficult for many Americans to buy into the idea of completely removing religion from public life. Crafty politicians may thus use the public outrage of secularists to establish their own faith-based bona fides. In other words, relentless separationist opposition to Christian nation advocacy can yield precisely the outcomes that secularists most fear. It is understandable that they were incensed by McCain's actions. It was a double provocation: the senator was claiming that the country was established as a Christian nation (provocation 1), yet he did so while clearly not believing that to be true (provocation 2).

What would have been a better course of action in response to these provocations? An organized secular movement might have coordinated a different type of attack. First, it would have identified its preferred GOP nominee (in the fall of 2007 that candidate would mostly likely have been Rudy Giuliani). Then it would have directed its operatives to complain about provocation 2, not provocation 1. In so doing, the movement might have led evangelicals to confirm their own worst hunches about the senator from Arizona. The prospects of the former mayor of New York could have improved greatly.

It is, obviously, easier to craft strategy with the benefit of hindsight. Yet the point stands: separationist interventions are often counterproductive. The Christian Right in the 1980s filled their fundraising coffers by invoking the specter of secular humanism. Revivalist politicians nowadays can use separationist advocacy to whip up their base. The anger of separationists needs to be properly managed by secular leadership. And if this movement could come off as a bit less separationist-y, it might help secularism's chances as well.

Pseudo-Accommodationism

Not all proponents of increasing Christianity's role in American public life are as cynical as John McCain. Many activists truly believe that more Christianity in our public life would benefit us all. And some of them, remarkably (and cynically), advocate this position while insisting they support separation of church and state!

One such example is Pastor Rick Warren, one of the most prominent evangelical leaders in America today. Warren refrains from Christian nation banter. In fact, he specifically disavows that he has any theocratic intent at all. Speaking at the Washington National Cathedral in 2008, he commented,

> I believe in the separation of church and state, completely. I do not believe in the separation of religion and politics. You can't separate them. I believe in the separation of church and state because I believe wherever you have state-sponsored religion, it dies. And a good example of that is where I just was, in Europe. In many places, the church has died because of the government sponsorship. I do not believe in a theocracy. I believe in pluralism. I believe in free market society. And I believe in a free market, may the best idea win. I do not believe in coercion, but I do believe in persuasion.[37]

Warren reiterated similar sentiments again for a huge national audience as he hosted the Saddleback Civil Forum on the Presidency. There he intoned, "We believe in the separation of church and state, but we do not believe in the separation of faith and politics, because faith is just a world view, and everybody has some kind of world view. It's important to know what they are."[38]

These remarks demonstrate the degree to which the term *separation* has become meaningless. Almost anyone, from a president setting up an office in the White House to dole out millions of dollars to faith-based social-service groups to a celebrity pastor, can now claim to be a principled defender of "separation."

Equally remarkable is the context in which Warren made this proclamation. The Saddleback gathering on August 16, 2008, marked the first time that the newly crowned Republican and Democratic nominees for highest office shared a podium.[39] That McCain and Obama agreed to stage such a momentous occasion at an evangelical megachurch, where they were subject to interrogation by a clergyman, is yet another testament to the plight of secularism.

As for Pastor Warren's comment on separation, it is one of his standard talking points.[40] The distinction he draws is illogical, not to mention contrived. Warren seems to be saying that since religious beliefs invariably influence a politician's worldview, we should know what those beliefs are. But that principle would seem to collide with the idea of separation that he claims to endorse. If Warren really does believe in separation, why does he want to know about a candidate's faith?

After all, a separationist believes that a member of Congress is not supposed to legislate on the basis of his or her religious scruples. Further, what good does it do the cause of separation (or the democratic process) to have clerics grilling elected officials as to their faith convictions? It seems more accurate to say that Warren doesn't believe in separation of church and state *or* separation of faith and politics. His rhetoric is profoundly disingenuous.

With that said, it is instructive to speculate as to what type of religion Warren would like to see in government and what his true motives might be. Do Warren and other high-profile conservative evangelical leaders really want more "faith," in a generic sense, in politics and government? Or do they want to elevate *their particular faith*? Revivalists often speak in soothing accommodationist tones. That is, they claim that a religion-friendly government, a government open to all religions, is their goal and desire. We might, however, justifiably counter that their true position is closer to *pseudo*-accommodationism.

Let's conduct a thought experiment: Could Pastor Warren tolerate a Mormon president? a Sikh attorney general? a Muslim director of Homeland Security? a Hindu secretary of state? a Bahai attorney general? How about three Wiccan justices on the U.S. Supreme Court, thrown in for good measure? In the name of bringing faith into the public sphere, could Pastor Warren countenance that cabinet and those appointees?

That's doubtful. It's doubtful because Mr. Warren is – to invoke Locke – orthodox to himself. He is convinced, understandably, that his faith is the most appropriate faith for guiding a nation under God. His endorsement of faith in politics is, however, misleading. Nothing in Mr. Warren's past indicates that he is an ultraliberal theologian who believes all creeds are equally good in God's eyes.

A troubling story about Warren made the rounds a few years back. It concerned a Jewish woman in Denver who asked him if she, as a Jew, would be saved.[41] To which Warren answered in the negative – a response indicating that her next stop would be hell. Christopher Hitchens saw the tale as a sign that Warren was "a vulgar huckster."[42] Yet Warren was merely being true to the rather exclusionary worldview that characterizes proselytizing Revivalists.[43]

Warren's own creed, incidentally, construes life in starkly capitalist terms. Warren and others understand something about evangelicalism's relative power position vis-à-vis other American religions. They understand that the "free market" he extolled earlier will permit evangelicalism to prevail over all competitors. They understand that the wall of separation is eroding. They understand that the super-well-oiled evangelical political machine has demographic assets, infrastructure, and networks in place that will bring its particular faith into American government faster than any other faith could possibly get there.

Indeed, this is why secularism is such a nemesis for Revivalists. It is not the movement's size or potential for growth. Warren is acutely aware of a truism we have noted here. "The actual number of seculars in the world," he taunted, "is actually quite small outside of Europe and Manhattan."[44] Rather, it is the judicial husk of separationism that is the problem. Those pesky midcentury legal restrictions prevent Christians from gaining access to new markets (that is, schools, other public spaces) where they might persuade others to come to Christ. All Revivalists need is for secularism to give up the ghost, and they will be ready to evangelize the nation. Accommodationism is the Trojan horse that will permit them to breach those markets.

Whether Warren and his colleagues would go the next step and install a Christian establishment is an open question. Yet it must be stressed that Christian Revivalists would *never* stand for true accommodationism because it would threaten their monopoly on the religious market. We are reminded of the brouhaha in the Minnesota state legislature back in 2001, when the Dalai Lama addressed the assembly. In the words of one outraged representative, "As a Christian, I am offended that we would have the Dalai Lama come and speak to a joint meeting of our Minnesota Legislature . . . He claims to be a god-king, a leader of the Buddha religion, which historically has been considered a cult because of its anti-Biblical teachings concerning the one true Holy God, Creator of Heaven and earth and His Son, Jesus Christ."[45]

This incident raises a curious possibility: were secularists to pursue a genuine accommodationism – whereby the government treats all religions equally – it might have the paradoxical effect of leading Christian conservatives to pine for separation! Accommodationism may have a legitimate secular application – lessening the Christian Right's eagerness to have faith, or their idea of it, play a role in government.

Since we are thinking aloud, might it be worthwhile for secularists to *get out of Revivalism's way*? This would mean that rather than contesting Revivalists, secularists would let them dominate the national conversation. It's a risky strategy and no quick fix, but it rests on two sound assumptions. First, the Revivalists are simply too extreme to govern. Their doctrinal obsessions would lead them to catastrophic decisions, internecine strife, and frightful overreach. Secularists must learn how to make rational use of Revivalist irrationality.

Second, although Americans in general might not be separationists, they are equally, if not more, averse to living under *someone else's* establishment.

The Threat of Establishment

Until now we have recommended a certain pragmatic passivity. Given the politically reduced circumstances in which secularism finds itself, there is much to recommend in conserving energy and avoiding needless confrontations.

There are, however, certain instances in which secularism must become proactive. It must be prepared to repulse *establishment threats*. Such threats involve *elected* public officials – not civilian pastors like Warren – who take tangible steps to establish Christianity as the national religion. Their advocacy leaves the realm of the rhetorical and demagogic and careens into the domain of the legislative. For secularists, this must issue an unambiguous call to arms.

The past decade has been rife with establishment threats. Let's start with the remarks of a member of Congress from Indiana. In 2004 Mark Souder offered this analysis of how he and his co-religionists think about their service to the country: "To ask me to check my Christian beliefs at the public door is to ask me to expel the Holy Spirit from my life when I serve as a congressman, and that I will not do. Either I am a Christian or I am not. Either I reflect His glory or I do not."[46]

Souder subsequently resigned his office in an adultery scandal.[47] But this is not why we should question whether he was fit for office. Rather, it is his willful misunderstanding of his role as a public servant that merits our concern. Souder was elected to defend the Constitution, not Christ. His constituents need him to answer to them, not God. His field of action is this world, not the next. As for the Holy Spirit, the representative from Indiana is enjoined to keep that under wraps in his capacity as a public servant.

The danger that Souder's legislative ethos creates is clear: if Souder and a bunch of other like-minded representatives refuse to check their beliefs at the door of public office, we are in trouble. If those others happen to be conservative Protestants (exponents of such views almost always are), it seems plausible that many in the House will not be restrained from establishing their version of Christianity in all matters of public policy. This establishment threat, then, consists of a caucus of like-minded evangelical and fundamentalist lawmakers who, through their concerted action, would impose their religious convictions on the American people.

Souder is not alone. Remarks from members of Congress about this being a Christian nation are increasingly commonplace. For example, in 2010 the Republican Stephen Fincher of Tennessee observed on the campaign trail for the congressional seat that he eventually won, "This is a Christian nation, and the road map to success as a nation is our Constitution and our faith."[48]

Similar thoughts were expressed that year by John Fleming, a Republican congressman from Louisiana: "We are either going to go down the socialist road and become like western Europe and create, I guess, really a godless society, an atheist society. Or we're going to continue down the other pathway where we believe in freedom of speech, individual liberties, and that we remain a Christian nation."[49]

Michele Bachmann of Minnesota strikes these themes as well. She once recoiled – as did seemingly half the Republican Party – from a comment President Obama made in Turkey: "we do not consider ourselves a Christian nation or a Jewish nation or a Muslim nation; we consider ourselves a nation of citizens who are bound by ideals and a set of values."[50] One can almost hear Bachmann, a Tea Party stalwart, shaking her head in anger as she demurs, "Saying we are not a Christian nation, that we are merely citizens with shared values. That is not true . . . Our founders said clearly that our Constitution was meant for a religious people. They were not ashamed of their faith."[51]

These types of rhetorical gestures emanating from members of Congress have been matched in recent years by congressional resolutions and bills to advance the Christian nation initiative. One notable effort is House Resolution 2104, known as the Public Prayer Protection Act of 2007.[52] The bill begins promisingly by noting that the First Amendment prohibits "federal establishment of any type of religious uniformity or orthodoxy that rewards observers and punishes violators."[53]

So far, so good. But then H.R. 2104 takes a peculiar turn. It declares, "*The First Amendment guarantees the right of elected and appointed officials to express their religious beliefs through public prayer.*"[54] From there it proceeds to state that "*the exercise of this right does not violate the Establishment Clause.*"[55]

And H.R. 2104 is still just warming up. In a fit of derring-do, the legislation sets its sights on the Supreme Court's ability to regulate the aforesaid public prayer: "*Notwithstanding any other provision of this chapter, the Supreme Court shall not have jurisdiction to review, by appeal, writ of certiorari, or otherwise, any matter that relates to the alleged establishment of religion involving an entity of the Federal Government or a State or local government.*"[56] H.R. 2104 is quite literally a threat to the Establishment Clause: it denies the judicial branch the right to rule on establishment cases! Sponsored by fifty-four members of Congress, all of them Republicans, the resolution was referred to committee and no further action was taken.[57]

Similarly, House Resolution 121, put forth in 2009, urged the president to recognize 2010 as "the National Year of the Bible."[58] The sprawling House Resolution 397 of 2010 attempts to "designate the first week in May as 'America's Spiritual Heritage Week' for the appreciation of and education on America's history of religious faith."[59]

The resolution rambles on, often sounding like the crib notes for a counselor at an evangelical summer camp. Yet for all of its emphasis on "spiritual heritage," only the Bible is mentioned. No reference is made to any religion except Christianity. The problem with these pseudo-accommodationist bills is that "spiritual heritage," "religion," and "faith" are euphemisms for evangelical Christianity.[60]

Resolutions such as these abound in the congressional record. They are largely ceremonial. Typically their sponsors trot them out with fanfare, knowing that they will ultimately be referred to committee and tabled. All the better to embarrass an "ungodly" Democrat, come election day! Yet secularists need to realize how firm a foundation has been set for future congressional classes to come around to the Christian nation perspective.

The actions of these groups are meant to align the government's policies with the worldview of certain Christian denominations. They comprise a very real and credible establishment threat.

Disestablishmentarianism and Accommodation

Let's say a Frosty the Snowman wearing a Christmas wreath and holding a cross (in his little branch hand) appears in the front yard of a municipal building. Separationists swing into action. After notifying the police, the ACLU, and AU – and, if possible, the Department of Homeland Security – they conduct their own independent inquiry. That investigation reveals that federal workers in the building had formed a Christian prayer group. One fine December day, they donned mittens and hats and gamboled down the stairs of the post office to erect this impromptu monument to their Lord and Savior.

Separationists, with their long tradition of crèche activism, know exactly how to respond. Truth be told, secularists have been responding to such provocations for decades. Symbolic micro-breaches of the wall of separation always send them to the courts or the op-ed pages.

A separation threat is not an establishment threat. This point needs emphasis. Separation threats are worrisome and annoying. But establishment threats, like the congressional resolutions nixing parts of the First Amendment as noted earlier, are potentially devastating to everything we hold dear about America. The coffers of secularism are bare, its personnel is limited. The whole movement needs to set its priorities in order to respond to the challenges coming its way. Separation threats simply should not be treated as lines in the sand when Revivalists have fanned out across all three branches of government and nearly control one political party.

Might it lead to an establishment? That is the question that a besieged, understaffed secularist movement should ask about state and federal policies. In the era of the Revival this is a more pragmatic prompt than *Does it entangle the state with the church?* Despite the dangers of rallying people around a nine-syllable word, secularists should, for now, move away from separationism toward *disestablishmentarianism.*

The latter is related to, but not identical with, separationism. One of the major differences is the willingness to at least *consider* the merits of accommodationist arrangements. According to this view, the American government has constitutional warrant to offer nondiscriminatory assistance to religion in general. As long as federal or state authorities favor no one religion in particular, they may dispense monies or material support or symbolic recognition to faith-based groups and maintain good civic conscience.

This does not mean, by the way, that the Constitution undeniably grants that warrant. Many legal experts have expressed convincing arguments against the idea that the Founders were accommodationists. As Leonard Levy put it, "The clause meant to [the Constitution's] framers and ratifiers that there should be no government aid for religion, whether for all religions or one church."[61]

This may indeed be true. But let there be no doubt, accommodationist principles are *already* being enacted by federal and state governments.[62] The Office of Faith-Based and Neighborhood Partnerships is a reminder of this. Statehouses around the country are entering into more and more partnerships with religious groups.[63] The Supreme Court, for its part, presently retains a majority that seems not at all averse to accommodationist logic. Secularists might not like accommodation, but it is becoming a fact on the ground.

Without making a virtue of necessity, there may be a rationale for experimenting with this approach. If a government offered truly equal support to all religious groups *and* nonbelievers, perhaps accommodationism could be fashioned into a workable alternative to separationism for secularists. There is a distinct advantage to this shift. The new approach is religion-friendly in the extreme. Were secularism to step back from the wall of separation, it might, in theory, gain new adherents. And there is nothing secularism needs more than that.

III

REVIVING AMERICAN SECULARISM

10

Who *Could* Be a Secularist?

> If a new coalition is going to succeed . . . it will have to include millions of theistic moderates, as well as a lot of people more like me, who consider themselves atheists, agnostics, or "spiritual but not religious."
>
> – Jeffrey Stout, *"The Folly of Secularism"*

In the word cloud that has emerged from our analysis so far, certain phrases about secularism's plight pop up conspicuously. These terms of distress include "People Problem," "Don't Do 'We,'" and "Chronically Lacking Numbers." Secularists can rant and rave all they want about the state of America today. But it'll all be hymns and psalmody to a Revivalist's ears unless they achieve one crucial goal: getting millions upon millions of their compatriots to share their core values and convictions. Reviving secularism is that simple. And it's that difficult.

Secularism needs numbers. The old midcentury approach isn't working anymore. The era in which a small minority could advance its agenda because it had sympathizers in at least one branch of government has long passed. The conservative Roberts Court ensures that it is no longer feasible to impose a "judicial strategy" on a restive population of Bible thumpers. Congress is at best apathetic to and at worst hostile to separationist secularism.

As for the White House's receptivity to separationism, think of President Obama's performance at the National Prayer Breakfast in February 2011. When a Democrat delivers a twenty-two-minute address about his personal faith, drops half a dozen scriptural references along the way, and declaims, "I came to know Jesus Christ for myself and embrace Him as my lord and savior"—all we can say is that *the sixties are over, man!*[1]

Under these trying circumstances, the proper strategy should be to make more secularists out of the American people. This excellent idea is somewhat hampered by the American people's lack of enthusiasm for secularism. In fairness, however, most Americans don't possess a deep understanding of what it means to be a secularist. And perhaps this leads us to one of the principle causes of the numbers problem.

Nowadays, the typical self-identified secularist is an atheist and a total separationist. (That this person is usually white, male, and extremely well educated is a point that we will leave unattended for now.)[2] This is a perfectly reasonable and respectable way of being secular. Atheist total separationists comprise an important secular constituency.

Yet this constituency is not synonymous with secularism as a whole. This subject's rich history is simply too full of complexity and possibilities to limit its personnel to this tiny cohort. Nonbelievers, in addition, are a tiny and disliked minority. Separationist initiatives, despite their many virtues, are not very popular. All of this adds up to the continuing marginal status of American secularism.

As a first step toward alleviating secularism's chronic shortage of supporters, let's advance a big-tent definition of what it means to be a secularist: *A secularist may be a believer or a nonbeliever, secular and/or secularish, but always someone who rejects religious establishment. She or he maintains that a good society is one whose government permits its citizens the maximal possible degree of freedom of religion and freedom from religion while maintaining order. A secularist is flexible as to how the government may accomplish this goal and is thereby willing to consider options ranging from separationism to accommodationism.*

Armed with this more inclusive definition, secularists might break into new markets. What types of people make up those markets? Let us imagine a continuum called the secular spectrum. Its poles are marked TOTAL SEPARATIONISM and ACCOMMODATIONISM. Many people situated within this spectrum don't understand or won't acknowledge their affinities with secularism. Others have peeled off from the pack because of the serious failure of secular vision and leadership, among other things. Ironically, many of the new markets are actually old markets that secular leadership was not able to consolidate or cultivate.

We know who *is* a secularist in twenty-first-century America. In the hopes of creating a future unmarred by depressing word clouds, let us pose the following question: who *could* be a secularist?

Christians: Spiritual Separationists

Not to overstate the case, but in theory, each and every one of the nation's 173 million or so Christian adults has a theological affinity with secularism.[3] For secularism is actually the spawn of Christian political philosophy.[4] Without Christianity it is hard to imagine how secularism would have ever come into existence. This doesn't mean that "true" Christianity is secular – there are a thousand ways to be Christian. It does mean, however, that very obvious genetic links exist, and these make alliances possible.

The parent, however, has disinherited the child. Christian antisecularism runs rampant in this country. One commentator complains that Catholics "are only half awake to the menace secularism constitutes for their religion."[5] In 2002, a former president of the Southern Baptist Convention warned that "the secular world is attempting to marginalize and demonize conservative evangelical Christianity."[6] In 2011 a Pew Forum survey of more than two thousand evangelical Protestant leaders around the world found that 71 percent "see the influence of secularism as a major threat to evangelical Christianity in the countries where they live."[7]

Leaving aside the question of what the term *secularism* meant to those leaders, it is a fact worth stressing that more than a few forms of Christianity lie on the secular spectrum. Many Christians, obviously, will never concede to that idea. It will take years of education and activism to convince others. But given the size of the market, and the preexisting links, the endeavor seems worthwhile.

How could the case be built? One excellent place to start is the Bible. Those activists delivering the secular gospel to Christians have some truly ace biblical proof texts at their disposal. Jesus's counsel to "render therefore unto Caesar the things which are Caesar's; and unto God the things that are God's" takes a disinterested view of worldly power, one that we examined earlier.[8]

We have already considered the astonishing advice of Romans 13:1–3, which advocates a respectful acquiescence to rulers. "Let every person be subject to the governing authorities" counsels a famous text in the Christian scriptural canon.[9] It goes on to advise that "whoever resists authority resists what God has appointed."[10] This counsel is apiece with 1 Peter 2:13–14 and 1 Timothy 2:1. A separationist tradition of apathy and even suspicion in regard to politics has a longstanding Christian heritage.

These texts point to the possibility that scads of flesh-and-blood Christians could be situated on the spectrum. Earlier we pointed to the Baptists in the revolutionary period who were among Jefferson's staunchest allies in his battle against state and federal establishments. It was the Massachusetts clergyman Isaac Backus who thundered, "Religion is ever a matter between God and individuals, and . . . no man or men can impose any religious test without invading the essential prerogatives of our Lord Jesus Christ."[11] During the early republic, groups such as the Unitarians supported disestablishmentarian causes. They would be joined by Universalists, Presbyterians, Methodists, Mennonites, and the Society of Friends, among others.[12]

We have already discussed the anti-Catholic "separationism" of nineteenth-century Protestant nativists. In terms of hard separationism, mainline Protestants in the mid-twentieth century served as the leading group within Americans United for Separation of Church and State (where some anti-Catholic impulses could once be detected as well).[13]

The affinity of mainliners for secularism is well known. It is often forgotten, however, that ultraconservative Christians once passively drifted in secular streams. In 1965 a certain man of God declaimed, "Preachers are not called to be politicians but to be soul winners . . . I feel that we need to get off the streets and back into the pulpits and into the prayer rooms."[14] This man of God was none other than Jerry Falwell, described as "the most visible American Evangelical of the twentieth century."[15] The segregationist was trying to get his fellow Protestant clergy to stop advocating for civil rights – a cause he suspected was being promoted by Communists and Satan.[16] That religion and politics should be separated was, for this fundamentalist, the gospel truth.

As is well known, Falwell had a change of heart later in life. This shift would hasten the Revival and lead to the inundation of the public sphere with fundamentalist and evangelical worldviews. But years before he founded the Moral Majority in 1979 and altered the course of American history, Reverend Falwell fell back on an age-old Christian intuition: true Christians refrain from politics.

For what it's worth, Christians today often come to that very conclusion. Late in life, no less a serial trespasser on church-state lines than Billy Graham walked back his enthusiasm for saturating the public square with (his) faith. Whether he was riffing with his buddy President Nixon about the "synagogue of Satan" or working with other Protestant leaders to derail the presidential bid of John F. Kennedy, Graham enjoyed a lifetime's worth of meddling in politics.[17]

But just a little while back, the famed preacher was asked by the magazine *Christianity Today,* "If you could, would you go back and do anything differently?" To this he responded, "I also would have steered clear of politics. I'm grateful for the opportunities God gave me to minister to people in high places; people in power have spiritual and personal needs like everyone else, and often they have no one to talk to. But looking back I know I sometimes crossed the line, and I wouldn't do that now."[18]

The Liberal Faiths

Secularists ought not hold their breath waiting for today's Christian conservatives to come around to Jerry Falwell's way of seeing things circa 1965 or Billy Graham's circa 2011. But they should remember this: even a heavily credentialed Revivalist *can* experience a secular revelation. Should that miracle occur, secularists *must* keep the door open for their new (old, actually) allies to walk on through.

There is, however, a huge column of religious Americans who have already experienced this revelation. Tens of millions of secular-friendly people are present in the ranks of Protestantism, Catholicism, and Judaism.[19] We shall refer to them as devotees of the "liberal faiths." Most could easily find a niche on our secular spectrum and most share the secularish qualities that we shall soon be discussing. One would be hard-pressed to identify a larger or more congenial market for the expansion of secularism.[20]

Once again, Protestantism is central to this story. It starts roughly near the end of the eighteenth century, mostly in Germany and the United States. It was there that theologians and clergy slowly developed alternative conceptions of God, humanity, scripture, ethics, science, and so forth. A leading historian of this development suggests that in the United States, liberal faith arose most prominently among Unitarians.[21] These Christians had grown disenchanted with the hell-and-brimstone puritanical Calvinism that had achieved a stronghold in the nascent republic. The alternative they sought, in the opinion of some, may be one of the taproots of the American liberal tradition itself.[22]

Scholars refer to this movement as liberal Protestantism, or sometimes liberal theology (this term is more specific to the Protestant milieu than the wider designation of liberal faith).[23] Its impact on American history is considerable. Liberal theology was a leading force in antislavery campaigns, the struggle for women's rights, the social gospel programs of the early twentieth century, and the (more religiously diverse) civil rights movement.[24]

During that same stretch of time, religions such as Catholicism and Judaism were also developing alternatives to their respective orthodoxies (to what degree this is due to the direct stimulus of Protestantism is not a question to be engaged here).[25] The contemporary United States is positively teeming with liberal faiths. These groups are large, organized, institutionally sophisticated, and, regrettably, pretty much estranged from secularism. An effective secular leadership will devise a strategy for reaching this audience. An effective secular leadership will make sure that members of this group become *a major part of the secular leadership.*

The liberal faiths share much in common, not least of which is the relentless throttling they receive at the hands of Revivalists. In 2010 the former Fox News personality Glenn Beck ridiculed Christian churches committed to a social justice agenda. He likened them, naturally, to Nazis and Communists, and in so doing he was engaging in a well-known smear.[26] Revivalists, as we have seen, often insist that they are orthodox while expostulating that all others are heretics. Often they accuse religious liberals of being secularists, by which they mean, of course, godless reprobates.

It goes without saying that all sorts of internal disagreements exist between and among the various liberal faiths. Yet the points of accord in matters political and theological are considerable. Those who have adopted the views of liberal theology see it as a "culturally transformative theology."[27] Social justice is a major component of its mission. As for politics, *in very general terms,* we could say that there is broad consensus on domestic issues such as guns (too many of them), environmental regulations (too few of them), gays (God bless 'em), NPR (ditto), abortion (a difficult and painful decision but ultimately the choice of the mother), evolution (most likely true), intelligent design (false and definitely not a science), poverty (among our nation's most pressing concerns), and so forth.

The one persistent fissure in the liberal religious camp centers on foreign policy. Reform and Conservative Jews, for example, often find themselves beating back divestment initiatives or resolutions critical of Israel advanced by pro-Palestinian Presbyterians, Episcopalians, or Evangelical Lutherans, among others. Still, aside from this highly divisive issue, there is considerable agreement.[28]

Yet the *theological* similarities among such groups provide the real opening for the future growth of secularism. The liberal faiths incline toward the spectrum of secularism for a variety of reasons. First, they place great emphasis on the individual's sovereign choice in reaching God.[29] We arrive at the divine through our own agency, reason, and volition.[30] External sources of authority, such as fixed interpretations of the Bible or religious tradition, valuable as they may be, can sometimes stunt an authentic spiritual journey, according to their view.[31] As for the federal government, it can neither aid nor prohibit our communion with the divine.

Second, today's liberal faiths are not hostile to spiritual diversity and equality. Multiplicity of sects is a good thing! The existence of many ways of contemplating and revering the divine does not vex liberals in the manner that it unnerves Revivalists.[32] Nor does the fact that these faiths are all equal in the eyes of the law.

Liberal theology, at its best, welcomes spiritual heterogeneity.[33] A state establishment of religion would do violence to this respect for diversity and for the equality of all. If asked to choose between a Christian nation and a nation under God, a Christian member of a liberal faith would most likely choose the latter. Accommodationism may have growth potential here. A government that encourages all religious groups to prosper would be congenial to a liberal theological worldview that values difference.

Which liberal theological groups represent a natural constituency for secularism? We could count Progressive Catholics, who emerged in the late 1960s buoyed by the liberalizing reforms of the Second Vatican Council. American Judaism has a deep historical connection with secularism that is evident among the so-called liberal denominations, otherwise known as the Conservative, Reform, Reconstructionist, and Secular Humanist forms of Judaism.

Among Protestants a few of the classic liberal denominations would be Unitarians and Universalists; the United Church of Christ and some other Congregationalist groups; and certain Methodists, Presbyterians, Episcopalians, and Evangelical Lutherans, among others. All of the groups mentioned have many disagreements, both between and among themselves. Still, the numbers they represent are truly staggering. Ecumenical bodies such as the National Council of Churches of Christ, as one example, have 45 million members![34]

Even though they fall naturally on the secular spectrum, the liberal faiths often have an uneasy relation with separationist secularism. Earlier versions of liberal Protestantism, for example, were virulently anti-secular.[35] To this day, members of the liberal faiths sometimes perceive secularism as militantly antireligious (the equation they have in mind is *secularist = extreme atheist*). In fact, liberal religious groups have historically found themselves occupying an uncomfortable "third way," or "mediationist," position, stuck somewhere between orthodoxy and infidelity.[36] Secular activism will need to rectify that problem by finding ways to let the liberal faiths comfortably situate themselves on the spectrum.

Like secularism itself, liberal faiths have had a difficult time of it in the late twentieth century.[37] It is not our place to explain why their numbers are leveling off or declining, but one reason needs to be mentioned. American Revivalists have had scores of effective, albeit extremely controversial and divisive leaders. Liberal faiths, by contrast, have lacked in this regard. From the 1970s forward no progressive figure has arisen as a counterweight to rainmakers like Billy Graham, Jerry Falwell, Pat Robertson, James Dobson, and dozens of others, right up to contemporary phenoms like Pastor Rick Warren. The assassination of Dr. Martin Luther King Jr. killed more than just the man and the dream; it left in its wake a huge void of ecumenical leadership and coordination in the liberal domain.

Secularists and liberal faiths have a mutual interest in huddling up. Both have been ransacked by the Revival. The latter has, however, developed in such extreme directions that an intriguing new reality is forming. The Reform Jew and the Unitarian share more in common both politically and *theologically* with each other (and the moderate nonbeliever) than they do with their ultraconservative brethren. The basis for an alliance is obvious.

The commonalities among the liberal faiths might extend in even more directions. Recent years have seen a marked increase in interfaith marriages in the United States. The authors of the recent book *American Grace: How Religion Divides and Unites Us* observe that "roughly *half* of all married Americans today are married to someone who came originally from a different religious tradition."[38]

Researchers such as Erika B. Seamon have noted that interfaith marriages usually include people who do not hold conservative religious views.[39] Such Americans, she adds, "can be classified as religious, spiritual, *and* secular. They embrace religion and spirituality in only certain areas of their life."[40] Once again, a shrewd secular leadership could make inroads among these secularish intermarrieds.

Religious Minorities: Pragmatic Separationists

We have looked at examples of religious Americans who embrace secularism for *theological* reasons. Their creeds enjoin them to keep government and religion at a distance from each other. Most religious minorities, by contrast, usually support secularism for *political* reasons. That is to say, their political interests motivate them to advocate for separationism.

It is easy enough to imagine the scenarios that minorities might dread in a nation with an official or unofficial establishment. They do not want to be second-class citizens. They do not want to encounter harassment on the job or discrimination in hiring. They do not want their children subjected to religious dogma, be it manifest or tacit, in public schools or other public spaces. They do not want subtle and not-so-subtle state-sanctioned endorsement of a religion that is not their own.

Religious minorities are natural allies of secularism and will often invest great energy in its defense. The bar has been set quite high by Jewish Americans who have performed countless heroic deeds on secularism's behalf. In the unlikely event that our government would erect a hall of fame in honor of those who have fought on behalf of the Establishment Clause, members of the tribe would merit their very own exhibit. As a writer in the conservative magazine *Commentary* declared (and lamented) in 1966, "The Jews are probably more devoted than anyone else in America to the separation of Church and State."[41] For most of the twentieth century the first commandment and mega civic mitzvah of American Judaism was *thou shalt keep the church apart from the state.*

By the middle of the twentieth century, as secular Jewish activism came into its own, a confident and mobilized community began aggressively to tout "the separationist agenda."[42] "Ever since the end of World War II," writes one scholar, "American Jews and the organizational structures that represent their interests have been at the forefront of organized efforts to influence the church-state jurisprudence of the Supreme Court."[43] Three powerful national organizations – the

American Jewish Congress, the American Jewish Committee, and the Anti-Defamation League—carefully monitored First Amendment questions. The agencies often cooperated, always bickered, and on occasion tried to disembowel one another behind the scenes.[44] But when all was said and done, their cumulative efforts helped bring about landmark decisions in the history of American secularism.

With the American Jewish Congress leading the charge, the three groups filed countless amicus curiae briefs, monitored developments nationwide, crafted complex legal strategies, and often brought to trial groundbreaking constitutional cases. One of the major figures in this movement was Leo Pfeffer. A rare combination of scholar and strategist, Pfeffer repeatedly appeared before the U.S. Supreme Court for nearly a quarter-century.[45]

Pfeffer, for example, filed an amicus brief in the *Engel v. Vitale* case of 1962. The verdict rendered by Justice Black about school prayer was a triumph for American secularism: "We think that by using its public school system to encourage recitation of the Regents' prayer, the State of New York has adopted a practice wholly inconsistent with the Establishment Clause."[46] Years earlier Pfeffer was to state the mantra that inspired him throughout his distinguished career: "Complete separation of church and state is best for church and best for state, and secures freedom for both."[47]

The Jews, at a scant 2 percent of the population, demonstrate how the passion and intelligence of small groups can embellish secularism. Countless religious minorities share positions similar to those of their Jewish compatriots. In a similar vein, a Muslim advocacy group recently bristled at a California mayor's suggestion that his city was "growing a Christian community."[48] "No government official or entity," wrote the group's leader, "should be in the business of promoting or favoring any one specific religion from their official position as an elected public official. We have a secular government and a pluralistic nation whose Constitution respects the practice of religion (or lack of it for those who choose to)."[49]

The Hindu American Foundation noted in a 2004 policy brief that "the separation of church and state must be maintained and that prayers should not be mandated in public schools during instructional time."[50] A statement by a Sikh American group strikes the secular theme as well: "The judiciary has long upheld the Constitutional guarantee of free exercise of religion and separation of church and state . . . It is troubling when people of faith are calling for the courts themselves to break down the wall separating church and state that has protected the right to freely exercise their beliefs."[51]

There can be meaningful understanding between religious minorities and nonbelievers under the umbrella of secularism, as was pointed out by the Dalai Lama in a speech in Tokyo in 2006: "Secularism does not mean rejection of all religions. It means respect for all religions and human beings including non-believers."[52]

Once again, religious minorities present secular activists with immense opportunities for coalition building. Effective secular leadership can swell (and diversify) the movement's numbers quickly by reaching out to these groups. This will involve listening to their concerns and quickly incorporating them into the movement's governing structure.

Can Anyone Who Isn't a Liberal Democrat Be a Secularist?

Many observers of the American political scene have crafted their own equation and it goes like this: *secularism = liberalism.* Drill down a little further and the formula expands to *secularism = liberalism = the Democratic Party.*

In the interests of increasing the size of their movement, secularists need to invalidate that equation. To pilfer a line from Barack Obama, secular outreach ought not be about red states or blue states, Republicans or Democrats, liberals or conservatives, but about serving all secularists in the United States of America (wild applause!).[53] There are large numbers of secular-friendly people outside the Democratic Party and outside liberalism. There are real hazards for secularism if secularists refuse to recognize that.

It is undeniable that there are profound and meaningful historical links between secularism and liberalism. Both *isms* emphasize individual rights, privacy, public reason grounded in science and rationalism, and untrammeled freedom of expression. Most secularists in

America view themselves as liberal Democrats, and vice versa. Still, in this quest for numbers a few caveats are in order.

First, secularism is *smaller* than liberalism. Secularism is something of a specialist. Its beat is religion and religion's relation to government. Secularism is there to tell you if granting tax-exempt status to faith-based lobbying groups is a good idea. Secularism can advise you as to whether that public school principal was out of line when he solemnly intoned "and now let us praise Him" at eighth-grade commencement exercises. Yet when it comes to foreign policy, or the economy, or international trade, secularism's opinions – which would, undoubtedly, be eloquent and witty – are far less substantive.

Liberalism is far *busier* than secularism. Liberalism has numerous and diverse policy prescriptions. These concern markets and diplomacy and trade tariffs and rules of war and the benefits of a multiparty versus a two-party system, and so forth. Those topics don't usually fall within the purview of secularism. Maybe they should. But the fact remains that secularism's interests are more specific. This means that not every good or bad idea associated with liberalism is necessarily essential to secularism.

Naturalization for illegal immigrants might strike some liberals as a very good idea. Secularism, however, reserves judgment and focuses solely on the identity of the population in question. If these potential citizens are wont to end up in a Revivalist worship center that views the United States as a Christian nation, then secularism – no matter what noble things liberalism might have to say about welcoming new Americans – demurs. Again, *secularism* and *liberalism* are not synonyms and this is why the reflexive equation that makes them identical is always problematic.

Secularism is inadvertently dwarfed by liberalism. Yet it is also being cold-shouldered, not inadvertently, by the Democratic Party. We have noted how the Democrats "got" religion after 2004. One result of their shift from separationism to accommodationism has been the diminished presence of concerns related to secularism in the party's rhetoric and policies.

With "friends" like the Democratic Party, perhaps secularists should, at the very least, explore the possibility of working with the "enemy." After all, the Party of Lincoln still shelters a few unlikely constituencies. Log Cabin Republicans, for example, are firmly ensconced within the GOP. Members of this gay and lesbian group have somehow established a niche for themselves among those who champion the Defense of Marriage Act and don't ask, don't tell legislation. If an organization whose members have run afoul of Republican policies could persevere in this environment, why couldn't secularists?

It's not as if secularism and the GOP don't share a little history. As we saw in previous chapters, Republicans in the nineteenth and twentieth centuries were great defenders of separationism. Admittedly, Revivalists have flooded today's iteration of the party. Still, groups who abhor Revivalist views can pick and choose, buffet style, from GOP policies that they feel match their own interests.

The case of libertarians – a group that sits squarely on the secular spectrum – is instructive, reminding secularists how a political movement can keep its feet in two rivers. The Libertarian Party's website reasons, "We favor the freedom to engage in or abstain from any religious activities that do not violate the rights of others."[54] There is nothing in the previous statement that would contradict any principle of secularism adumbrated in this book.

Indeed, one libertarian scholar advances the argument that "social harmony is enhanced by removing religion from the sphere of politics."[55] Another points out that "libertarians are not for or against religion; they oppose government policies that favor religion, in part because this means government must define what constitutes a religion. This will inevitably favor the status quo at the expense of smaller, newer religions, and at the expense of individual liberty."[56]

There are, however, some natural conflicts. Libertarians evince a thoroughgoing suspicion of the state. In classical secular theory, by contrast, the state is the guarantor of individual religious liberties and social order. This having been said, libertarianism, self-described as "fiscally conservative and socially liberal," seems to have a helpful message for future secular activism.[57] It blurs timeworn political categories, a crucial concept because emerging voting formations, especially those of younger generations, may move beyond the old red state, blue state distinction.[58] Likewise, future secular activism might strategize beyond the polarities of Left and Right. The hard-to-pigeonhole libertarians offer an instructive model: a revived secularism should be equally difficult to categorize.[59]

That having been said, it would be silly for secularists to ignore their affinity groups within classic liberal and Democratic precincts. A few natural allies come to mind. Gay Americans have an interest in a sort of "free exercise" approach to consensual sexuality. That is to say, they don't want Congress or any other government entity limiting their personal or sexual freedom. They have a corresponding passion to prohibit an establishment of compulsory heterosexuality.

In truth, secularists could learn much from the history and practice of gay activism in the United States. The recent legalization of same-sex marriage in New York is a testimony to the dividends of slow, patient, well-organized, and well-funded advocacy.[60] Gay Americans have run afoul of Revivalist religious institutions, with dire consequences for their civil rights. They have an affinity with the secular movement's goals, and a robust gay-secular alliance is crucial to secularism's future.

Finally, we alluded earlier to certain gender imbalances within self-described secular movements in the United States. A concerted effort by a mostly male movement to reach out to women, and to let women *lead,* may quite possibly be the single most relevant coalition-building move that secularism could execute. Needless to say, there are many issues of interest to women, and they take many positions concerning them. But in terms of a natural fit, pro-choice activism makes perfect sense.

Many women have learned the hard way that Revivalists, be they evangelical Protestant or traditional Roman Catholic, would have no compunction about curtailing women's reproductive freedoms. If secularism, with its centuries-long commitment to the idea of privacy, cannot manage to lock arms with women and men interested in protecting these rights, it is not worthy of being a political movement. With *Roe v. Wade* in the cross hairs of mobilized Revivalists, the moment for concerted action by pro-choice feminists and secularists is now or, more likely, yesterday.

We should not, however, fall into the erroneous logic that all religious groups are anti-choice. As a NARAL Pro-Choice America document itself points out, many of the liberal faiths we surveyed earlier "have been at the forefront of the reproductive rights movement."[61] Let it be reiterated that religion is not the enemy of secularism and secularism is not the enemy of religion. The revival of secularism depends on the degree to which all involved can grasp that insight.

11

How to Be Secularish (In Praise of "Secular Jews" and "Cafeteria Catholics")

He recited the Mourners' Kaddish. Over a sinking coffin, even a nonbeliever needs some words to chant, and "*Yisgadal v'yiskadash* . . ." made more sense to him than "Rage against the dying of the light."

– PHILIP ROTH, *Zuckerman Unbound*

SECULAR HUMANISM, A SMALL but substantive school of philosophy, never garnered much attention, even among secularists.[1] Only Revivalists took its teachings to heart. In the 1980s fundamentalists and evangelicals worked themselves into a tizzy depicting secular humanism as a national security threat, a Communist menace, and the Antichrist all wrapped in one package.

Francis Schaeffer, whom we had the pleasure of meeting earlier, was one of the first conservative Christian intellectuals to sound the alarm. In his *Christian Manifesto* he fulminated, "The humanistic position is an exclusivist, closed system which shuts out all contending viewpoints . . . [it] is completely intolerant when it presents itself through the political institutions and especially through the schools."[2] With Schaeffer uncovering the "humanist conspiracy against the faith," his co-religionists followed suit.[3] Thus one very agitated Revivalist could write that "this new religion, SECULAR HUMANISM, is destroying the breaking system, casting off all restraint.

We are already on fire; SECULAR HUMANISM is throwing gasoline on the fire."[4]

Secular humanism, humanism, atheism, secularism: they were all evil and immense and – how many times have we seen this before? – they were all thought to be the same thing. In 1985 a *New York Times* writer joked that secular humanism stood for "everything they [the Religious Right] are opposed to, from atheism to the United Nations, from sex education to the theory of evolution to the writings of Hemingway and Hawthorne."[5]

What was it about secular humanism that so riled the Revivalists? Maybe they got it into their heads that this ungodly perspective had the Supreme Court's blessing. Chalk that misunderstanding up to Justice Hugo Black's infamous footnote 11 in the *Torcaso v. Watkins* case in 1961. There he casually referred to secular humanism as one of a handful of "religions" that do not "teach what would generally be considered a belief in God."[6] The logic of calling secular humanism a "religion" remains inscrutable to this day – especially to secular humanists.[7]

Yet, as the Revivalists marshaled in the 1970s and 1980s, they became convinced that something sinister was afoot. The U.S. government, they suspected, was trying to establish the religion of secular humanism.[8] Senators such as Orrin Hatch of Utah batted around amendments prohibiting federal funding for any educational programs that included secular humanism in their curriculum.

As Leo Pfeffer put it in his perceptive study of footnote 11, fundamentalists "suddenly realized that they had discovered a good thing and were not about to let it go."[9] Any curricular component they objected to – the teaching of evolution, sex education, literature with mature themes – could be tarred with this brush. Revivalists, we have seen, know how to reap legislative and fundraising bonanzas by exaggerating the beastliness of their enemies.

For the record, this enemy was anything but a beast. On the contrary, it was a tiny movement, obscure and academic in equal parts.[10] Secular humanism, as we know it today, is mostly associated with the humanist philosopher Paul Kurtz. The unbelievably prolific University of Buffalo professor – an entire book is needed to catalog all of his writings – set up his offices in Amherst, New York, in the sixties.[11] In 1969, he and his colleagues opened a small publishing house, Prometheus Books. It lovingly printed classic and new texts on nonbelief and skepticism.[12]

Kurtz's group had done much to keep the humanist flame burning. Its publishing arm, for example, reissued the *Humanist Manifesto,* originally published in 1933. That text, whose guiding spirit was the philosopher John Dewey, opens as follows: "The time has come for widespread recognition of the radical changes in religious beliefs throughout the modern world. The time is past for mere revision of traditional attitudes. Science and economic change have disrupted the old beliefs. Religions the world over are under the necessity of coming to terms with new conditions created by a vastly increased knowledge and experience."[13]

Science. Reason. Progress. The greater good. Such are the lofty earthbound columns of the humanistic worldview. Kurtz updated the 1933 statement with the *Humanist Manifesto II* of 1973. "The next century," that document boldly predicted, "can be and should be the humanistic century."[14] Another statement, the 1980 *Secular Humanist Declaration,* declaims, "We believe that it is possible to bring about a more humane world, one based upon the methods of reason and the principles of tolerance, compromise, and the negotiation of differences."[15]

The secular humanists, as the reader may have gathered, were fond of the manifesto genre. At the turn of the century they issued *Humanist Manifesto 2000: A Call for a New Planetary Humanism* (not be confused with *Humanist Manifesto III* of 2003).[16] Working the global angle (and failing to anticipate George W. Bush–style unilateralism), *HM 2000* called for "a bicameral legislature in the United Nations, with a World Parliament elected by the people, an income tax to help the underdeveloped countries, the end of the veto in the Security Council, an environmental agency, and a world court with powers of enforcement."[17] One can just imagine how today's Tea Partiers might cotton to the idea of paying an extra tax for the construction of women's health clinics in East Timor!

Kurtz, showing his debt to Holyoake, referred to "eupraxsophy" as the foundational idea of secular humanism. By this he meant "good practical wisdom" that "draws its basic principles and ethical values from science, ethics, and philosophy."[18] As often happens with highly educated people who cogitate about things like eupraxsophy, popular recognition eluded the philosopher.[19] Except among Revivalists, most Americans knew very little about secular humanism. Back in 1984, when Orrin Hatch was anathematizing it in Congress, the legendary Democratic senator Daniel Patrick Moynihan admitted, "I have no idea what secular humanism is. No one knows."[20]

The humanist century has gotten off to a rough start. As we noted earlier, Kurtz was deposed from the leadership of the very organizations he founded. Proponents of "angry atheism," he charged, had usurped the movement.[21] As if secularists didn't have enough to worry about, their long tradition of organizational self-implosion seems to be intact. The same antagonisms witnessed in the showdowns between Holyoake and Bradlaugh and the crackup of the National Liberal League have resurfaced. The script is well known: moderate bridge builders and pugilistic radicals lock horns, become divided, and are subsequently conquered.

It is too early to assess the legacy of secular humanism, but a few observations seem warranted. It was, undoubtedly, a serious school of thought that attracted quality thinkers. It nobly enshrined the skeptical tradition through its publishing division. It established stable, albeit small, organizational structures and research units in the United States and abroad. Perhaps its greatest contribution, however, has remained unnoticed. It was one of the few American movements that thought of secularism as lying *outside* a strictly political, legal, or (anti)theological framework.

American secular movements have always been narrowly focused on the proper relation between state and church or on nonbelief. Those are certainly not inconsequential concerns. But they don't offer a smidgen of insight as to how to *be* secular: how we might live our lives, or think about ethics, or interact with others.

Secular humanism tried to rectify that problem. It offered its members a new philosophy for thinking about the world. It wanted to create secular humanistic denizens of a secular humanistic planet.

Its leaders wanted you to come aboard, to become a secular humanist. "Let us work together," exhorted *Humanist Manifesto II*, "for a humane world by means commensurate with humane ends." "We invite others in all lands," it continues, "to join us in further developing and working for these goals."[22]

This endeavor is well intentioned but futile. A revived secularism should not try to engineer a new species of secular man. It should not try to lure individuals away from their present identities. It should not try to create a nation of secularism bound by ideological unity and flying a secular flag (what would such a flag look like, anyway?). It should not come across as a new religion bearing secular revelations. None of that will work because people already have a rich sense of identity, often centered on religious affiliation. Few of these people seem inclined to convert to a new state of existential being proposed by a school of philosophy.

Instead, the way forward consists of getting individuals to understand that being secular or secularish is *already a part of who they are.* A milestone will be reached when Americans "embrace the adjective" – when they start casually referring to themselves as secularish Catholics, secular lesbians, secularish Latinos, and so forth.

Getting people to embrace the adjective should not be prohibitively difficult (and it is nowhere near as challenging as the dead lift of personal-identity makeovers that the secular humanists attempted). Whereas secular humanism tried to elevate the species to the loftiest of heights, a more modest course of action is likelier to succeed. Millions upon millions of people are already living secularish lives. They are believers and nonbelievers, Protestants and Muslims, Republicans and Democrats, what have you. The goal should be helping people, *as they are* and *where they are,* to recognize that they *already* abide by core secular and secularish virtues.[23]

The Realm of the -ish

Let's get our bearings by probing the subtle but significant differences between secular and secularish individuals. This is an experimental and analytical distinction; it needn't be adopted by all secular people.

If it rejected the *-ish,* secularism would not have its corpse picked clean by some horrific flock of razor-beaked Revivalist birds. But the secularish category has certain practical and conceptual advantages. Practically, the quantity of secularish people out there is immense since they are found in so many different religious traditions. It would be nice to have a way to refer to them.

Conceptually, this designation helps us home in on those hard-to-describe nonpolitical dimensions of secular identity that figures like Holyoake and Ingersoll explored in their work. Earlier, we found that secular states may exert secondary "bonus" effects. These draw out from, or instill in, citizens certain viewpoints, ethical dispositions, and values. This is the realm of the *-ish,* which we will now visit, with the travel advisory that colloquial references to *secular* may overlap with *secularish* in possibly confusing ways.

When someone, martini in hand, mentions to you at a party, "I consider myself to be a secular Muslim," he or she is not necessarily saying "I'm a separationist, a dues-paying member of Americans United for Separation of Church and State, in fact." A different connotation seems to be in play. But what?

What is the difference between an American secular Muslim and an American secularish Muslim? The American secular Muslim is a person who supports secularism because he is in accord with a political philosophy about governance that prohibits an establishment of religion (in all likelihood this person's specific fear is being subject to a Christian establishment). The secular Muslim in America *might be very religiously conservative.* He might even agree with the Christian Revivalist on issues such as the role of women, homosexuality, artistic freedom, and so forth. But since this person is situated in the Christian United States, he does not want the federal government to endorse the ideas of a Christian church.

Of course, some conservative Muslims, like the Baptists we encountered earlier, might endorse separationism on *theological* grounds. That is to say, there may exist religious justifications for a Muslim to be wary of church-state alliances. In fact, a distinguished

Muslim jurist has recently argued just that. In a startling book Professor Abdullahi An-Na'im demonstrates that classical Islamic civilization itself has traditionally been far more receptive to disestablishmentarian politics than to theocratic politics![24]

Leaving that aside, let's enter the realm of *-ish*. The secular*ish* Muslim in America shares the secular Muslim's aversion to Christian Revivalist politics. Yet whereas the secular Muslim could very well be an orthodox Muslim, the secularish Muslim is decidedly unorthodox. Both think the same way about the relation of politics to religion. But the secularish Muslim practices and conceptualizes her religion differently. In this sense she joins all the other secularish in all other religious traditions. She has in common with the secularish Catholic and the secularish Hindu a variety of characteristics worth examining.

Let's start with *moderation in belief and practice*. If the secularish had a motto, it would be "Don't get overwrought." The secularish are serene about matters of faith. They confront grand metaphysical concerns with a temperate spirit.

Take, for example, the grandest of all grand metaphysical questions: does God exist? A secularish person, as we understand her, could conceivably offer any answer from the following possibilities: "Probably," "As far as I can tell," "I think so," "Can't be sure," "It could be," "It doesn't seem likely," "Probably not," "Who knows?" or "Could I possibly know such a thing?"

She wouldn't, by contrast, respond with the churlish refrain "That's so offensive. Of course He does. How dare you ask such a dumb question!" It's not that secularists are all fence sitters or closeted agnostics. Rather, they are people who are at peace with ambiguity and uncertainty as regards things that cannot be proved.

The secularish, however, are not opposed to speculating about grand metaphysical questions. The German philosopher Jürgen Habermas referred to this as "postmetaphysical thinking," or being "prepared to learn from religion while at the same time remaining agnostic."[25] Aside from the agnostic part (the people considered here run the gamut of theist and atheist perspectives), this quotation captures well the open-mindedness of the secularish.

The secularish are opposed to making the fruits of their metaphysical speculation matters of public policy (and here they earn the designation secular as well). Pondering the question *What would Jesus do?* can be enlightening and interesting. It is the type of discussion that should be pursued in lecture halls, houses of worship, seminaries, and elsewhere. But in a democratic society, where differing conceptions of Jesus abound (and where not everyone believes in Jesus), such a query cannot drive policy discussions.

A more intriguing question might be *What would Jesus do when confronted with the problem that even his followers completely disagree as to what Jesus would do?* That is a much better prompt, the type that secularish people love to investigate. Their bracelet would read WWJDWCWTPTEHFCDATWJWD.

Moderation is a function of the high threshold for doubt that the secularish enjoy. Doubt is moderation's personal trainer. *All* of us, even the most sincere believers, have doubts. Theologians often concur. They stress that expressing uncertainty about God is by no means unusual. "In the Protestant understanding," writes one authority, "doubt cannot be separated from faith. Serious doubt about God is possible only in reverence toward him, and reverence issues necessarily in a radical criticism of our beliefs and values."[26]

The willingness of Jews to entertain doubt is the stuff of legend. And fiction. In her famous short story "Bloodshed," the writer Cynthia Ozick depicts the reaction of a Hasidic rebbe who chides a secular Jew who has slipped into his prayer services:

> "Sometimes" the rebbe said, "even the rebbe does not believe. My father when he was the rebbe also sometimes did not believe. It is characteristic of believers sometimes not to believe. And it is characteristic of unbelievers sometimes to believe. Even you, Mister Bleilip – even you now and then believe in the Holy One, Blessed Be He? Even you now and then apprehend the Most High?"
>
> "No," Bleilip said; and then: "Yes."[27]

The fictional rebbe in this story expresses a misgiving about God's existence that is extremely well known among nonfictional Jews. Yet

Revivalists, like the rebbe, rarely permit that doubt to alter their religious *practice.* An ultra-orthodox Jew may on rare occasions question the existence of God, but those doubts infrequently are reflected in his actions. There is (occasional) doubt, but there is never moderation. He observes the Sabbath scrupulously. He adheres to rigid dietary laws. He worships in strict accordance with traditional Judaism's calendar. In this meticulous ritual observance, the ultra-orthodox Jew is joined by hundreds of millions of Revivalists in other traditions worldwide for whom there is no moderation in *practice.*

The secularish are precisely the opposite. They are the type of people who have no difficulty missing daily prayers. A secularish Jew can miss lots – conceivably a lifetime's worth – of Sabbath services, as a secularish Catholic might avoid Mass. A secularish Muslim might enjoy herself a good strong drink every day after work. Coffee might be the jump-start to a secularish Mormon's morning routine.

Christian Revivalists and other conservative religious groups are ostensibly not at peace with this. Most clergy in these quarters would prefer that their charges take (the clergy's interpretation of) religion more seriously than anything else in their lives. More prayers should be mouthed. More alms should be given. More rules should be obeyed. More time should be spent contemplating the divine. Faith must be the priority, spiritual perfection the goal.

Let the truly pious pursue that goal. As we have already noted, the truly pious may be secular (in the strict political sense) as well. But the endeavor of nonsecular folks to force *everyone* to participate in this quest constitutes an immense danger. Countries like Iran and Saudi Arabia offer grim models of the violence done to individuals and collective bodies when states try to "democratize" spiritual perfection. It was no less a lover of God than Augustine, incidentally, who reminded us, "Normally, both our knowledge of God's will, and our love of him, are in this life imperfect."[28]

And since we are on the subject, why must God and religion always be the first priority, anyway? Why must those two always take the gold? Why not a bronze medal, or maybe honorable mention, in that hierarchy of values by which a person orders his or her life?

What could be more important than God, you ask? Your children.

Your friends. Your parents. People outside your faith whom you love. Art. Music. Charity. The trees. New York City. Does the possibility exist that a good and gracious God might concur?[29] Secularish believers like their religion and love their God. They may be lax in the eyes of the orthodox – but for them that is a matter between them and their Creator.

As for nonbelievers, the same standards apply. Atheists must also incline to moderation. If they are extreme in their nonbelief, they are atheists but not *secularish* atheists. And maybe not even secular ones! Some nonbelievers, as we have seen, can be so intolerant of the very existence of the church that one wonders if they can be called secular – since secularism is a political philosophy about relations between *church* and state.

The adjective, we stress, connotes a strong distaste for metaphysical extremism of any kind. Most nonbelievers, of course, are deeply secularish. Atheists routinely celebrate Christmas with family members and can do so without delivering a tableside anti-Christian manifesto. Agnostic parents will send their children to a high-performing Catholic elementary school, all the while disagreeing with the teachings of the church. Secularish nonbelievers don't get overwrought.

This-Worldliness

The secularish are here-and-now people. They live for this world, not for the next. Of course, many religious secularists believe that there is, in fact, a hereafter. They may anticipate spending a restful eternity therein. While they are down here, however, they don't obsess over what lies ahead.

The theme of living for this world has an organic place in *both* religious and nonreligious thought. What sociologists call the "this-worldly" orientation is articulated in countless adages and aphorisms declaimed across the millennia.[30] Thus, the author of the book of Ecclesiastes brooded, "Only this, I have found, is a real good: that one should eat and drink and get pleasure with all the gains he makes under the sun, during the numbered days of life that God has given him; for that is his portion."[31]

In the modern frame we see this idea reflected in George Santayana's well-known quip: "There is no cure for birth and death save to enjoy the interval."[32] An analogous impulse led the French singer Georges Moustaki to muse, "*Nous avons toute la vie pour s'amuser. Nous avons toute la mort pour se reposer.*"[33]

Why are the secularish so focused on this life? As for nonbelievers, that's easy to answer: because they are certain that it is the only life that they are going to get! Atheists and agnostics accept their mortality and seek to make the most of the limited time they are granted here on earth. "We must live to the utmost degree," declares a character in Lara Vapnyar's *Memoirs of a Muse.* "We are atheists, so there is no afterlife for us. Thus, death is nothing."[34]

This ethos rankles certain religious critics. They contend that such a lazy acceptance of human finitude has repercussions for one's moral outlook. Living for this world, and this world alone, they allege, triggers an absence of probity as well as nihilism and depravity.

All of that is certainly a possibility. But conservative theist critics are too quick to draw a one-to-one correspondence between not believing in the afterlife and not believing that there is anything wrong with the practice of child sacrifice. This line of reasoning has become something of a cliché in the American culture wars. It invariably begins with a (mis)quote of Dostoyevsky's axiom from *The Brothers Karamazov,* to the effect that "if there is no God, everything is permitted."[35] The moralist in question asserts that if you don't fear God, if you don't think about hell or heaven every waking moment of the day, you will obviously permit yourself to engage in fiendish acts.[36]

But the logic is sumptuously false. Let us leave aside for now the long history of fiendish acts committed by believers. Being resigned to a finite existence does not automatically lead to immorality, contumely, and rapine. A glance at the history of movements advocating nonbelief would bear this point out. Freethinkers have rarely characterized themselves as libertines or reckless partiers embarking on a lifelong bender. On the contrary, these groups have tended to implore nonbelievers to advance a better, purer, more moral alternative to a world corrupted by *religion.*

A relentless ethos of do-goodery best describes how atheists and agnostics seek to behave. Back in 1871, in his *Principles of Secularism,* George Jacob Holyoake defined *secularism* as "making the service of others a duty of life."[37] *Humanist Manifesto II* advocates for the environment (pretty forward-looking for 1973), reminding us that "ecological damage, resource depletion, and excessive population growth must be checked by international concord."[38] Kurtz's *HM 2000* seeks "to ameliorate the human condition, advance happiness and freedom, and enhance human life for all people on this planet."[39] Does this sound like nihilism?

But they are not the only ones who validate the here and now as the arena for acts of moral virtuosity. Many religious traditions identify our world as divinely sanctioned. They thus concur with nonbelievers: this world *is* good and it is the stage upon which one *does* good.

We spoke earlier of a secular outreach to Christians. Secular activists could remind the latter about the Christian lineage of the word *secular* itself. The great Saint Augustine (354–430) spoke in his writings of something called the *saeculum.* The term is commonly translated as "the world" or "the age." The precise meaning of *saeculum* that the bishop of Hippo intended is subject to debate.[40] One historian renders it as "'existence'—the sum total of human existence as we experience it in the present, as we know it has been since the fall of Adam, and as we know it will continue until the Last Judgment."[41] In the words of another commentator, Augustine defended "a place for the secular within a religious, Christian interpretation of the world and of history."[42]

According to Augustine, we are all imprisoned in the *saeculum.* We "all live together" down here—the saints with the sinners, the saved with the damned.[43] This existence will come to an end only with the arrival of the messiah. Yet his entrance cannot be hastened or induced. He will come when he is good and ready to come. So here we are, "groaning together" (as Paul so eloquently put it in Romans 8:22) in secular, as opposed to heavenly, space.

One might say that a Christian has a certain respect for secular time; this is, after all, where God put her. Maybe it's not the best of all worlds. But, as one writer phrases it, Christian faith "*affirms* the world . . . because it is a world endowed by a Creator God with harmony, beauty, intelligibility, and usefulness that has not been entirely erased by the effects of human sin."[44]

The this-worldly spirit is well known in the modern liberal Jewish milieu through the concept of *tikkun olam,* "repair of the world." By the mid-twentieth century the term had come to mean that "we human beings (not just the rabbis) fix the world of concrete objects, animals, and persons, by engaging in both environmental and social care and repair."[45] The Jew's moral obligation is to improve the present world for others.

Secularism's this-worldly tincture has a policy implication. Secularism cannot tolerate religious groups that are *otherworldly* in radical ways. The most dangerous proponents of faith are those who abhor the *saeculum,* who view it as incorrigibly corrupt, sinful, and carnal. Their hatred of the present world leads them to *hasten* their (and our) promotion to the next. A leading and troubling indicator of an anti-secular worldview is an exaggerated concern with one's fate after death.

This is the psychological mechanism that permits the suicide bomber to ignore all those children he or she passes along the aisle of the bus. This is the justification of the Christian fundamentalist who wishes to trigger the apocalypse by setting off wars in the Middle East that conform to the plan laid out in the Scofield Reference Bible. At present, Islam has a far greater problem with these death cults than does any other faith. But from secularism's point of view, it is not a particular religion that is the problem. Instead, it is a theological "mode"—one that radically degrades this life and overvalues the next. That is what the secularist cannot tolerate.

Now that our government has established an Office of Faith-Based and Neighborhood Partnerships to help religions do good, it seems reasonable to set up a department that will monitor potential dangers related to the opposite possibility: that religions can sometimes do very bad things. This is a controversial claim, yet a secular state cannot ignore groups whose rhetoric and theology radically degrade the life we live together on earth. The state has an interest in keeping such groups under surveillance and proactively incarcerating their members when they constitute a threat. The government has a responsibility to curtail religious freedom when it threatens order (and our survival).

Theological Off-Roading and Facing Down Fatwas

Foremost among the secularish virtues is the ability to be critical of oneself and others.[46] Starting with the latter, one of secularism's great gifts to the species is the principle that we are endowed with the right to think anything that we please. Even the curmudgeonly Luther defended freedom of conscience and famously proclaimed thought to be free.[47] Jefferson, however, provided the epitome of a pull quote – one so good we invoke it a second time: "But it does me no injury for my neighbour to say there are twenty gods, or no god. It neither picks my pocket nor breaks my leg."[48]

Freedom of conscience is the gateway drug to the so-called expressive liberties. In 1859 John Stuart Mill, in his classic essay "On Liberty," identified the "inseparable" link between the architects' liberty of conscience and the eventual "liberty of expressing and publishing opinions."[49] Gaining these rights was a huge struggle for the English infidels, with their contrarian opinions. Holyoake, Bradlaugh, and countless other Victorians were tossed into British dungeons for their anti-Christian candor.[50] It is because of their heroism that secularism gained a new plank: the right to publicly criticize religion, or anything else, for that matter.

This insight needs a little modernization. In today's secular democracies the person who is exclusively critical of *others* does not necessarily qualify as secularish. Revivalists, after all, are always finding fault with those who are not like them. So are New Atheists. What the secularish person adds is a willingness to question *inwardly*. She is the one who thinks about her religion outside the established lines. She does a little "theological off-roading."

This does not mean that all secularish believers are heretics. It does imply that they ask lots of hard questions about their own faith. Everyone knows a Catholic who is perennially criticizing and disobeying her bishop. Everyone knows a "freestyle" Muslim who does things her own way. Everyone knows a Protestant who disagrees with his church's anti-gay teachings. The Jewish penchant for self-critique, whether it be disagreements with rabbis or the policies of the state of Israel, is legendary.

Most religious groups have made peace with the idea that others possess this right to critique. The French political scientist Olivier Roy has spoken of "shared rules of the game."[51] By this he refers to the agreement of believers not to resort to "violent or illegal forms of opposition" when confronted with criticisms or policies they do not like.[52] Most religious groups play by the rules of the secular compact. Yet extremist Muslims have difficulty accepting them. Not a month goes by without a filmmaker or novelist or cartoonist stumbling over the tripwire of their sensitivities.

Perhaps the defining incident of this kind was the fatwa placed on Salman Rushdie by Ayatollah Khomeini following the publication of *The Satanic Verses* in 1988. We need not rehearse that grim tale here.[53] What we should stress is that the pattern it established has become a distressing norm. Again and again we have seen the same script play out. An artist – Muslim or non-Muslim – says something unflattering or perceived to be unflattering about Islam. Extremist demagogues get wind of it and send their legions into the street. Rioting ensues. Embassies are besieged. Bookstores are firebombed. People are killed. Threats are made left and right.

And expressive freedoms are gradually dialed back. "Who would dare write a book like *The Satanic Verses* nowadays?" asks a journalist in an overview of the twenty years that have elapsed since its publication.[54] Who indeed?

Being secularish involves the conviction that criticism of self and others is important and necessary. It also involves strong support for governments that defend that right. Finally, it may eventually involve the actual physical defense of that right. One recurring charge aimed at secularists is that they make no sacrifices, they live for no ideal higher than themselves. For what cause, it is asked, would they suffer? Here is an apt response: secularish people must be willing to put themselves in harm's way for this critical freedom.

Cafeteria Catholics and Secularish Jews

Compared to those of secular humanism, our entrance requirements for membership among the secularish are undemanding, even a bit on the cheap side. They consist of broad mental outlooks that can be held while *simultaneously* clinging proudly to a Muslim or Catholic or Mormon or atheist identity. No doubt a stand-alone secular or secular humanist identity is feasible, but hybrids are preferable. Hence the power of the adjective. Being secular or secularish can complement and enhance a rich preexisting identity.[55]

Secularish people are everywhere. It is true, however, that certain cultures mint such individuals in larger quantities. Although it is not widely known and not statistically measurable, Catholicism is a veritable greenhouse for the secularish soul. It is exceedingly difficult to get Catholic leaders to acknowledge this fact. Fortuitously, we have the recent and remarkable speech of the Fordham theologian Tom Beaudoin to expose a glorious truth: "There are secular [it might be appropriate to add the *-ish*] Catholics in my family, there are probably some in yours. They constitute the oceanic and silent penumbra of the Catholic Church."[56]

For Beaudoin, "secular Catholics, by way of analogy to our brothers and sisters who are secular Jews, are those raised Catholic who cannot find Catholicism as their central life project. Secular Catholics are those baptized Catholics who find themselves having to deal with their Catholicism, and to do so as an irremediable aspect of their identity but whom 'we' in ministry and theology might be tempted to call 'nonpracticing,' 'religiously illiterate,' 'relativistic,' 'inactive,' or 'fallen away.'"[57] It is very clear from Beaudoin's remarks that these people, although often judged to be "cultural victims," are true Catholics, perhaps the silent majority. They are, he continues, self-scrutinizing to a fault.[58]

Which bring us to secular Jews.[59] *Secular Jew* is a common term that doesn't sound out-of-the-ordinary or self-contradictory. But consider the slight dissonance that occurs when you say "secular Muslim," or "secular Mormon," or "secular Southern Baptist." The ease with which Jews have taken up the secular mantle makes them distinct among faith groups in America. In fact, the distinction between secularish and secular Jews is almost nugatory; most Jews could be labeled as one or the other.[60]

Secular Jew is a familiar cultural trope, deeply ingrained in the popular imagination. Nearly everyone in the United States knows and loves and is perhaps even married to a secular Jew. A variety of images and even stereotypes are associated with this group: They are city dwellers holding advanced degrees. They are comedians and iconoclasts. They are your second-grade teacher, Mrs. Fantowicz. They are also dreaded First Amendment avengers. You wouldn't want to be caught on the lawn of the statehouse sliding the Baby Jesus into his manger in the company of a secular or secularish Jew.

To a degree unmatched by any other religious group, Judaism has embraced the adjective, owned the adjective. Consider that there is an actual *denomination* known as Secular Humanistic Judaism. It comes replete with rabbis, synagogues, seminaries, hymnals, and so forth.[61] SHJ is lively and intriguing. Yet it doesn't even account for a fraction of the many members of the tribe who are secularish.

A recent poll in the United States indicated that 44 percent of all "Jews by religion" surveyed referred to themselves as "secular" or "somewhat secular."[62] The authors of the study noted that no other religious group came anywhere close to that number. Buddhists placed a distant second. Only 22 percent were willing to affix the s-word to their identity.[63]

In fact the most audacious nationalist project of modern times, the Zionist movement, was born not of rabbis and Talmudic scholars, but of secular Jewish intellectuals. The pioneers, such as Theodor Herzl, Max Nordau, and Ze'ev Jabotinsky, who dreamed of a homeland for the star-crossed children of Israel were not guided by traditional Jewish law (*halakah*) nor by a call from God. Quite the contrary. The Jews who gave us Zionism found inspiration in Enlightenment reason, or socialism, or communism, or other ideologies birthed in modern Europe.[64]

Not all Jews who left (or, more accurately, fled) Europe in the late nineteenth and twentieth centuries went to the land that eventually became Israel. Most actually immigrated to the United States. They brought with them something known as Yiddishkeit. This word might best be translated as "Jewishness." It refers to a lifestyle built around the culture and language of Yiddish inherited from the old country, eastern and central Europe.[65]

Cresting in the mid-twentieth century, Yiddishkeit in America was at once profoundly Jewish *and* secular. It was intolerant of excessive religiosity. It abhorred displays of hyper-"fruminess," or by-the-book adherence to Jewish law. Nor did it show an exaggerated respect for the institution of the rabbinate. In its more communistic variants it was aggressively anti-theistic. Yet for the most part these secular Jews were fiercely proud to be Jewish Americans. They were Jews in moderate and creative ways that did not, understandably, always make the rabbis proud. As one secular Jew put it at midcentury, for Jews "secularism is not synonymous with anti-religion."[66]

This culture's, this *Jewish* culture's interests were wholly worldly: poetry, politics, knishes, literature, baseball, labor activism, music, social justice, free speech, hot dogs, theater, Franklin Delano Roosevelt, public education, basketball, comedy, and so forth. It left a remarkable imprint on American Judaism, and on America itself. Anyone who fancies Broadway show tunes, or belongs to a union, or fancies bagels, or calls someone a schmuck can say a little thank-you to secular Yiddishkeit.

Or perhaps say a little prayer. As goes the world, so goes Judaism – the incandescent secular and secularish cultures of Yiddishkeit and secular Zionism have dimmed in recent decades. The global Revival certainly did not pass Judaism by; on the contrary, it has ransacked these cultures in both the United States and Israel. Secular and secularish Judaism, conceivably the most successful hybrid of modern values and religious culture ever produced, is currently fighting what one sympathizer calls "a defensive, rearguard battle."[67]

Secular Judaism will, however, likely make it through intact. At present its proponents are making a stand against the condescension and political heft of their Revivalist brethren. The following comment made by a leading Israeli proponent of secularism encapsulates the plucky resolve of these Jews: "My only claim is that I, who do not observe the commandments, represent traditional Judaism as much as a Jew who observes the commandments and sometimes even more so."[68]

Yet non-Jewish forms of secularish identity will also need a boost. The goal of the secular movement is to make the world safe for other secularists who embrace the adjective, be they secular or secularish in thought and practice. And it is for this task that we must now plan.

12

Tough Love for American Secularism

> Secularism is neutral. It is neither a dogma nor a doctrine. If anything, it's an abstention. Secularism abstains from favouring one religion over another, or favouring atheism over religious belief. It is a political principle that aims at guaranteeing the largest possible coexistence of various freedoms.
>
> – AGNÈS POIRIER, *"The Pope's Plot Against Secularism"*

AN AD EXECUTIVE TASKED with devising marketing materials for the secular movement could do far worse than this slogan: "Secularism: Freedom from Religions You Really Don't Like!" Although daft, this mantra does capture one of the breakthrough intuitions of the architects of secularism: no one should be subject to the faith of another.

Luther, Locke, Williams, Jefferson, and Madison had a world-beater historical insight there. Nowadays that insight goes over swimmingly with religious minorities, nonbelievers, libertarians, intermarrieds, lots of gay folks, and certain types of Christians, among others. Revivalists, however, are far less impressed by this logic.

They assert that the Founders did not intend for us to be free from (Protestant) Christianity. The First Amendment, they stress, was worried about the establishment of a particular denomination of Christianity, not (Protestant) Christianity itself. There is not, allege the Revivalists, any constitutional right to freedom *from* our nation's official Christian religion. Every now and then a non-Revivalist meets them halfway, such as when Senator Joseph Lieberman posited the idea that "the Constitution promises freedom of religion, not freedom from religion."[1]

But maybe even conservative Christians in America can draw inspiration from our new PR slogan. Revivalists feel that *they* are the ones subject to the "faith" of another. The American government, they protest, is not neutral. The religion it preaches might be called secular humanism or secularism or liberalism. Call it whatever you want, but it's a religion nonetheless. This sinister creed elevates science over faith, approves of aberrant lifestyles, disrespects unborn life, and espouses other sundry abominations. An existing establishment of secularism in the United States, they complain, denies Christians their basic freedoms.[2]

One can break this argument down fairly easily, and we will return to this theme later. Experience, however, suggests that the person who believes that this is, or should be, a Christian nation will continue to believe this no matter what arguments and evidence are adduced. Little suggests that these impassioned advocates will let something like logic, fairness, or the common good stand in their way. Which leaves secularists scratching their heads as to how to protect the secular state from being overrun by Revivalists. If the previous chapters demonstrate anything, it is that conservative Christians are making reasonable headway toward this goal. What is the best way to respond?

In classical secular theory, force is always an option. The state, Locke taught us, has authority over all religious groups who threaten social order. Then again, it is mostly a moot point: American Revivalists have almost never threatened social order. True, on rare occasions they have lost their marbles. At the first Republican presidential debate of the 2012 campaign, the founder and chairman of the Faith and Freedom Coalition, Ralph Reed, reeled off a litany of Revivalist complaints about the nation. He then reminded his listeners that the Declaration of Independence authorized Americans to "[replace] the government by force."[3] This comment was, naturally, disturbing. But it was out of the ordinary as well.

For, as we have seen many times, Revivalists don't need to resort to violence. They are doing just swell by working through democratic processes. They have built a social and political movement of formidable strength and depth. They have mobilized voters, fronted candidates on the state and local levels, and emerged as a lobbying juggernaut in Washington. Their leaders have trained a class of intellectuals to sway public opinion on crucial issues. Perhaps most important, conservative Christians have become an indispensable component of the Republican Party's base. It is not unwarranted to claim that they have become part of the leadership of the GOP itself.

Every one of these activities is perfectly legal and perfectly in accord with democratic principles. And taken as a whole, those activities spell out secularism's doomsday scenario: the gradual infiltration and takeover of the secular state by Revivalists. Every fight-or-flight bell in the secularist steeple is chimed. Will such a state remain secular for long? Will religious minorities be relegated to second-class citizenship? Will nonbelievers be subject to discrimination? Will the civil rights of women be curtailed? Will homosexuality be criminalized? What will happen to secularish worldviews? Will a breakdown of social order ensue?

The greatest threat to American secularism is not the Muslim extremist trying to nuke Times Square – though that too is something to lose sleep over. Rather, it is the *lawful,* and triumphant, march of the Christian Right down the boulevard of liberal democracy. To be a secularist today is to get continually trounced in democratic arenas, be they courthouses, legislatures, or public opinion polls. To be a secularist today is to be on the defensive, outnumbered, out-hustled, outfoxed, and ensnared in the midst of a Great Awakening; only God knows when its flames will be extinguished.

"Hello, my name is secularism and I consistently get outgeneraled in the public sphere by Revivalist groups – maybe I'm not so invulnerable as I was once led to believe. I guess I thought too well of myself. Maybe I should have reached out more to others."

Hello to you, secularism, thank you for sharing, and maybe we should stop freighting you with unreasonable expectations. The secular idea, it must be stressed, is not a prescription for a social utopia. It was, is, and always will be *the least bad alternative for achieving peace in complex, religiously pluralistic societies.* If someone wishes to propose a better program for managing such societies, she is more than welcome to try. Good luck with that. But in the interim, it is worth sticking with secularism, with its somewhat seedy past and all.

The Revival of Secularism: A Twelve-Step Program

So what's the plan? Secularism must confront some hard truths. In order to achieve long-term goals it will need to check into rehab. Once there, it will be asked to experiment, change strategies, compromise, listen to others, and endure living arrangements that may be less than ideal. What follows is a series of recommendations, a twelve-step program for the survival and revival of secularism.

1. Know Thyself: This book began by pointing to a debilitating crisis of definition. This crisis not only renders public discussions of (and assaults on) secularism absurdly imprecise, but also impairs its ability to function as an instrument of change. The primary casualty of the confusion surrounding secularism's definition is secularism itself.

To refresh your memory, here is the definition that can move secularism forward: Secularism is a political philosophy that, at its core, is preoccupied with, and often deeply suspicious of, any and all relations between government and religion. It translates that preoccupation into various strategies of governance, all of which seek to balance two necessities: the individual citizen's need for freedom of, or freedom from, religion and the state's need to maintain order.

Of course, not all secularists agree as to what is the best strategy for regulating the relation between government and religion. It is for this reason that we have rejected the equation *secularism = separation*. Strict separation of church and state is only one possibility for achieving the aforesaid balance.

Other equations were rejected as well. *Secularism = atheism,* as we have seen, is a late-blooming derivative approach. The explicitly political definition of secularism, which developed in the high-speed thought corridor stretching from the Reformation to the Enlightenment, is preferable and far superior to the theological (or, more accurately, anti-theological) definition that emerged in the nineteenth-century branch of infidel Victorian thought (that is, the definition that Holyoake contested). The latter entangles secularism in precisely the metaphysical scrums that are irrelevant and possibly hazardous to the project of maintaining order in heterogeneous societies. It is time for secularists to read Holyoake again.

As it moves forward, secularism will need to be distanced from other associations. It is not, of course, akin to Stalinism (though it must study some of the genetic affinities and extract valuable lessons). Nor is it perfectly accurate to say that *secularism = liberalism.* As for the old standby, *secularism = anything the Democratic Party does,* it should be rethought in the age of Obama.

2. Admit That the People Problem Is Real and Serious: Chapter 1 spoke of "Locke's escape clause" – an often unnoticed feature of the secular basic package. It stipulates, somewhat paradoxically, that democracy must be preserved undemocratically.

In other words, there will be no establishment of religion *even if the majority wants one.* To the unhappy masses who want more faith in public life, secularism says, "Keeping your religion out of government, schools, and other public spaces is better for all of you. We know this from experience – y'all should read about it yourselves sometime. Now please return to your houses of worship, quietly."

This reply reflects the situation today, but how long will that last? Many formidable legal scholars are presently dismantling the theoretical infrastructure of the secular state. Sooner or later, one of them will figure out how to convince a court to pronounce the escape clause illogical, undemocratic, and even tyrannical.

We also pointed to a clear preference for individual rights over collective rights in the thought of the architects, a tradition that reached its zenith in the writings of John Stuart Mill. Secularism's (and liberalism's) adulation of the individual can be politically debilitating – especially when its adversaries are hardwired for communal electoral action.

In a way, the vaunted judicial strategy of the twentieth-century separationists is a perfect symptom of secularism's underlying anthropophobia, or social anxiety. To refresh your memory, this approach enabled separationists to achieve stunning legal gains without having to win over too many American hearts and minds – a people-free solution to the threat of Protestant establishment!

The limitations of this strategy, however, became apparent when the composition of the U.S. Supreme Court reddened in a more conservative direction. At the same time, legal scholars were beginning to critically reassess the classical claims of separationism. And while all that was going on, Revivalists were seeding courts and law schools with their own homegrown talent.

The Lockean escape clause is about to be annulled. The liberal celebration of the individual is not conducive to forging communities that win elections. The judicial strategy is no longer reaping benefits. This means that secularism needs to change course. It does not necessarily need a statistical majority to achieve its goals. But it will need millions more people who, in some way, support its core beliefs.

The operative word here is *people*. People vote for members of Congress. People run for Congress. People sit on PTA boards. People raise money for social causes. People stand up to Revivalists. Secularism needs people.

3. Turn to Religious Moderates for Help: As secularists ponder expanding their movement to markets where there are people, two things about the potential clientele should be considered.

First, most of them are religious people who are alarmed by extremism, including Christian nation politics and other Revivalist provocations. Many of these people fall under the rubric of liberal faith and are secularish. Some can be designated as religious minorities. However, some minorities in the United States (Muslim Americans come to mind) may be far more conservative in their outlook. They are secular, not secularish. For political reasons they too are ready to take up the adjective. It is secularism's job to make sure the movement's program is open and accommodating to them.

Second, none of these religious people, be they secular or secularish, want to come to secularism in the way Christians come to Christ. They have no desire to become secular humanists or Brights or New Atheists. Their reasons for joining a secular coalition will be pragmatic, self-interested, and focused exclusively on preserving their own religious liberties and political freedoms. They'll just take the adjective, thank you. And that's just fine. Every religious person who embraces the adjective makes American secularism stronger.

4. Understand the Value (and Etiquette) of Coalitions: Secularists could learn a lot from Revivalist leaders. Falwell's Moral Majority, for example, was not denominationally specific. The same held true for its offspring, the Christian Coalition, and today's evangelical movement.

Evangelicalism operates via a "para-denominational model." It culls people from *across* different Protestant denominations. One observer describes it as "a loose affiliation (coalition, network, mosaic, patchwork, family) of mostly Protestant Christians of many orthodox . . . denominations."[4] Evangelicalism is a mode of worship that unites people through simple creeds (for example, the infallibility and inerrancy of the Bible, the centrality of accepting Christ as one's personal savior) and clear political positions.

The principle of co-belligerency permitted Revivalists to make alliances with groups that shared some of their political concerns, but not necessarily all of them. What is fascinating about the Christian Right is that in pursuit of political gains it managed to put aside profound blood-soaked theological disagreements. Conservative evangelicals, for example, were somehow able to work with traditionalist Catholics, Mormons, and ultra-orthodox Jews on a range of issues pertaining to abortion, gay rights, public schools, and so forth.

The secular leadership of tomorrow will need to be something of an aggregator. It too will need to bring together many diverse groups in a broad coalition. Coalitions, however, demand a certain etiquette. Rule number one for the participants is *Do not lambaste your fellow coalition members*. This is why the more extreme atheist groups, with their penchant for mocking all believers, are so hazardous to secularism's already failing health.

One Reform rabbi in Britain describes religious moderates in his country as feeling as though they have been "stabbed in the back, then turned around and punched in the face" by militant atheists.[5] Another commentator expresses his frustrations with one leading New Atheist: "My real problem with [Richard] Dawkins is that he wants to pick a fight with me but he doesn't. He picks a fight with a kind of literalism, or fundamentalism, that – I would hope – most of us do not hold. He ignores centuries, even millennia of religious sophistication in Judaism and liberal Christianity and equates religion with a plain reading of the Bible."[6]

The secular movement has to secularize itself. In other words, there are to be no public discussions of metaphysics, theology, or God's existence or lack thereof among the members of the coalition in their capacity as members of the coalition (unless these occur in a bar, after hours – but even then . . .). Let us refer to this as Holyoake's First Principle of Secular Thermodynamics: to maintain peaceful equilibrium in the coalition, theist and anti-theist talking points must cancel themselves out, resulting in a productive, unifying silence on speculative matters.

If conservative Catholics and evangelical Protestants could put aside half a millennium of enmity in the name of making abortion illegal, there is no reason why a nonbeliever and a Sikh American (who have no particular history of antagonism) couldn't work together to put a stop to that prayer circle "spontaneously" breaking out in the cafeteria of Middle School 234.

Who will coordinate the many different players in this grander, more inclusive new secular coalition? The vanguard group will need to appreciate diversity and be diverse itself. After all, this alliance may, to borrow a quote from Woody Allen, look like the cast of a Fellini movie. It will include everyone from pagans to Baptists to gay Americans to feminists to libertarians to Muslim Americans to secularish Christians to Reform Jews to Scottish Rite Masons, and so many more. The leadership must reflect that heterogeneity as well.

Religious moderates (as this book keeps insisting) are one of the largest, most financially solvent, and most organizationally sophisticated of all potential coalition members. These moderates – this can't be repeated enough – must be part of the leadership structure itself. The vanguard will need to deputize an entire "ministry to the moderates" and that ministry should be staffed by moderates.

This book has mentioned the drawbacks of letting extreme atheist groups speak on the behalf of secularism. There are, however, secular groups that are not built on atheist premises. We have encountered one of these already: Americans United for Separation of Church and State (AU). Now under the capable leadership of the Reverend Barry Lynn, it has fought valiantly on behalf of secularism for more than six decades.[7] With a minister of Christ at its helm, the organization has the right idea and is presently secularism's greatest hope. AU, however, is separationist in its politics, and strict separationism may be an impediment to recruiting certain potential members.

5. Don't Forget the Catholics: The uneasy relation between Catholicism and secularism has been touched upon often. The difficulties began at the beginning, so to speak. Luther, after all, was a person whose hatred for the papacy was so unbounded that he referred to it as "the scarlet whore of Babylon."[8] Protestant-Catholic animosities have festered frequently and often tragically, and the complex triangulating role that secularism has played in these antagonisms has yet to be seriously studied.

We saw, for example, how anti-Catholic separationism found a home among nineteenth-century Protestant politicians, whether Republicans or xenophobic nativists. (In the early twentieth century scholars have detected anti-Catholic separationism among American and British leftist intellectuals as well.)[9] The wall of separation became a convenient means of discriminating against Catholics.

In fairness, we should add that the Catholic Church has often maintained its own set of unhelpful phobias about secularism.[10] One key document that exemplifies this tendency is the "Syllabus of Errors" of 1864. An observer notes that these collected teachings of the church were interpreted "by liberals everywhere . . . as a declaration of War."[11] The syllabus presents a litany of the newfangled mid-nineteenth-century ideas that Rome emphatically rejected. This extreme statement ridiculed even the suggestion that the church might not be superior to the state. Specifically, it denounced the idea that the "Church should be separated from the state, and the state from the Church," calling it an error.[12] Of course, by the time of the Second Vatican Council in the 1960s, Catholicism had moved away from positions like this one.[13]

Antagonisms between secular Protestants and Catholics flared up again in the mid-twentieth century, though this time around Jews got into the mix as well. Many of the epic separationist battles of the 1950s and 1960s, especially those involving public aid to parochial schools, pitted Protestants and Jews against Catholics.[14] The Catholics sought public funding for their sectarian schools. Protestants had their usual reservations about that. Jews, whose ranks were filled with public school teachers, had their own reasons for opposing that arrangement. This created no small amount of Jewish-Catholic tension. For example, critics of Leo Pfeffer, the separationist mastermind, accused him of being an anti-Catholic bigot.[15]

Secular leadership needs to resolve this tension with Catholicism. It has already been suggested that someone (but who?) should apologize on behalf of secularism for the nineteenth-century injustices we encountered earlier. The best option here is to listen attentively to what secularish Catholics – whose ranks are legion – have to say about secularism and act upon their recommendations.

On political grounds, Catholics simply cannot be ignored. They comprise the largest single religious denominational voting bloc in this country, accounting for 25 percent of the population and 30 percent of the vote.[16] Catholics are notoriously difficult to pigeonhole, sometimes voting Democratic, sometimes voting Republican. Their electoral behavior routinely beguiles pollsters and party operatives.

But most important, Catholic voters beguile their own bishops. Although secular worldviews are highly unpopular in the church hierarchy, this doesn't necessarily mean that lay Catholics are opposed to them. Time and again, Catholics have shown themselves to be truly independent voters.[17] One of secularism's crucial long-term projects consists of joining in common cause with secularish Catholics.

6. Articulate a Clear, Simple Mission and Creed (Emphasis on Simple): By now it is clear that secularism has been paralyzed by its proponents' inability to define it. Now that a specific definition of secularism has been advanced and scores of pages devoted to its exposition, perhaps it is time for us to look at a more concise, market-friendly description of secularism. Marketing materials, recruitment efforts, team colors, ring tones, and stationery logos should promote one core idea: *disestablishment.*

There are good reasons for staking the future of secularism on disestablishmentarianism. First, unlike separation of church and state, it has unambiguous constitutional sanction. Originalists cannot deny that the Founders prohibited an establishment of religion.

True, they can, and have, claimed that nonestablishment applied only to the federal government, not to the states. In fact, one commentator suggests that the goal of Christian Right activists is to seek "a return to a time when state legislatures and courts could create a patchwork quilt of personal liberties and individual freedoms."[18] Let the Revivalists go ahead and try to repeal the Fourteenth Amendment, thereby letting statehouses run roughshod over the civil liberties of all.[19] Secularism likes those odds.

In any case, another advantage is the commonsense appeal that nonestablishment holds for many religious Americans. As our ad executive has intuited, *very few people want to live under someone else's establishment.* Even among the many groups in the conservative Christian coalition, this would create a problem. In other words, how happy would a Lutheran of the Missouri Synod be living under a Southern Baptist establishment, or vice versa?

Traditionalist Catholics and evangelicals often team up on political issues, but any student of religious history is acutely aware that, even today, neither would tolerate being subject to the religious authority of the other. Evangelicals comprise roughly a quarter of the population.[20] There is no reason to assume that the other three-quarters of the country want to be subject to their dominion, even if it is pitched broadly in feel-good Christian nation terms.

Many equations cited so far have been in some way insufficient, but *secularism = disestablishmentarianism* sounds promising. It could encompass multiple strategies: accommodationism, separationism, variants of the *laïcité* approach, noncognizance (to be discussed in a moment), and so forth. Nothing, incidentally, prevents a shrewd secular leadership from simultaneously deploying all of the strategies just mentioned, on different fronts, in an effort to advance its agenda.

7. Shelve Separationism (for Now): Secularism must repeat to itself the following mantra: reasonable political movements set reasonable political goals. Total separation is not, at present, a reasonable goal. This is because American secularism's fundamental objective, the movement's central priority at present, is impossible to achieve. Secularists must recall that politics is the art of the possible.

Total separation of church and state is a nonstarter in the White House and it matters little if its occupant is a Democrat or a Republican. The Supreme Court's long-term pattern of decisions in this arena suggests it is abandoning the conceptual vocabulary of mid-century secularism. Staple ideas such as "wall of separation," "pervasive sectarianism," and "*Lemon* test" are being toppled by the Left and the Right.[21] Even the right of individual taxpayers to bring establishment cases has been, as we have seen, rescinded. For a reminder of Congress's recent record when it comes to defending the wall, refer back to the House resolutions discussed in Chapter 9.

To restate a point made earlier, separationism is riddled with vulnerabilities. For starters there does not seem to be clear constitutional warrant for walling off. Next, the human-rights track record of separationist regimes has not been stellar. Any honest observer of history realizes that Britain, with its established Anglican Church, is and was a much better place to live than that separationist Shangri-la, the Soviet Union.

There is also the problem of what we might call the "inevitability of entanglement." It is impossible for the state and the church to have absolutely nothing to do with each other. Citizens will profess faith and the state will have to deal with that, whether that means providing military chaplains, or excusing federal workers for days of religious observance, or figuring out how to tax ecclesiastical properties. The example of France drove this point home: even in a society in which religion is viewed with profound skepticism and sometimes hostility, the government retains intimate, almost supervisory, relations with its communities of faith.

Finally, for whatever reasons, total separation does not seem to have broad popular appeal in the United States. It is not a warm and fuzzy concept on the order of liberty or freedom of speech. Many Americans find the approaches associated with activists like Michael Newdow to be extreme and absurd. Whether this lack of popularity is justified or not is an entirely different question.

8. Depict the Secular State as a Referee and Give Accommodationist Policies a Test Run: Revivalists, we noted, tend to see secularism as an establishment in and of itself. At this teachable moment let us make a constructive response: the ideal secular state is not an establishment, but a referee. Its job is not to cater to a particular religion's worldview, but to make sure that no single religious worldview gains unfair advantage over any other through exploiting the apparatus of the state (for purposes of fairness, atheism or secular humanism could be categorized, in this context, as a "religion" as well).[22]

Let us be clear: the state's primary responsibility is to referee *itself.* It must understand its own massive power and endeavor to carefully monitor how it treats its religious and nonreligious citizens. Religious liberty blossoms when there is no "collusion," or "match fixing" – when the state judiciously refuses to favor one team. This permits everyone, from atheists to Zoroastrians, to play.

The state can be referee-like in a variety of ways. One is to completely disregard religion and ensure that it has nothing to do with the state. In the name of fairness the government would vow to take no note of religion whatsoever. James Madison spoke of "non-cognizance," which one scholar describes as the idea that "religious citizens are to be treated the same as all other citizens, with no distinctions made on the basis of religious affiliation. Civil government is to be blind to religion as such."[23] This variant of separationism has some sterling qualities. It is simple and precise. Yet it suffers from all of the drawbacks that afflict separationism, which were enumerated earlier.

There is another way in which a state can be judicious in its capacity as arbiter. Accommodationism, as we saw, requires the state to exhibit scrupulous equanimity in dealing with all religious groups. A self-refereeing state would be fair to all religions by extending equal benefits to all. To quote one well-known statesman, "The secular state is a state which honours all faiths equally and gives them equal opportunities."[24]

This was the logic that animated the federal faith-based initiatives. The idea was that a government would use its resources to assist all religious groups and, in the process, reap benefits for the larger society—yet the results that the office has produced are cause for skepticism. Suffice it to say that many kinks need to be worked out.

Still, in the interests of widening its appeal, secularism should not rule out accommodationist approaches. There are obvious dangers here, and it bears repeating that this approach is an experiment. But it offers advantages as well, the biggest being that few religious groups in America could oppose it.

Which brings us to a second advantage: those who do oppose *authentic* accommodationism are often Revivalists. Here is an interesting sidebar to the saga we explored in Chapter 8. When President Bush first announced his plan to create an office for faith-based initiatives, both Pat Robertson and Jerry Falwell responded with criticism and suspicion. Both traditionalists were fearful that money would go to the "wrong religions." Robertson was concerned about Hare Krishnas, Falwell about Muslims (Robertson's group later received half a million dollars in faith-based grants).[25] Accommodationism, if practiced correctly, is a far bigger threat to Revivalists than it is to secularists. The former insist they simply want more "faith" in public life; one interesting aspect of an accommodationist secularism would be putting that dubious claim to the test.

9. Use the Extremism of Revivalists Against Revivalists: Revivalists are formidable. They are not, however, invincible. They are plagued by fairly obvious weaknesses that a shrewd secular leadership could exploit, often by just stepping out of the way. Self-confidence and unyielding belief in their own convictions have powered Revivalists to much political success, but these qualities have a flip side: chauvinism and moral arrogance. These attitudes are thunderously at odds with America's moderate and serene ecumenical environment that inclines toward toleration and pluralism.

Revivalist Christian groups in this country are fundamentally *not* tolerant. They are not Lockeans—nothing indicates that they see toleration as the "chief characteristical virtue" of Christianity. Completely orthodox unto themselves, they have far more difficulty with true religious diversity than their embrace of accommodationism indicates. Their impulses are missionizing; they actively seek to bring as many people to their side of the spiritual ledger as possible. All of these qualities are perfectly legitimate in the private sphere, but they tend to outrage Americans when they make the pilgrimage to public space.

It seems that every time a conservative Christian comes into the national spotlight, we experience that outrage. In 2007, commenting on a Republican presidential rival, Mitt Romney, Mike Huckabee asked an interviewer, "Don't Mormons believe that Jesus and the devil are brothers?"[26] The former governor of Arkansas later apologized to Romney for the remark.[27]

In the same election cycle John McCain was endorsed by two pastors, John Hagee and Rod Parsley. Within nanoseconds of their public statements to this effect, the press dug out remarks each had made indicating a less-than-healthy appreciation for the nation's glorious tradition of interfaith respect and peace. Parsley had averred that "Islam is an anti-Christ religion that intends through violence to conquer the world."[28]

Hagee, for his part, had a history of belting out some of the old standards of anti-Catholicism. He had once written that "Adolf Hitler and the Roman Catholic Church joined in a conspiracy to destroy the Jews."[29] McCain repudiated both endorsements in May 2008 after the pastors' statements were strongly criticized across the nation.[30]

In fact, anti-Catholicism runs rampant in Protestant Revivalist circles. In 2011, the Republican presidential candidate Michele Bachmann left her church, which was associated with the conservative Wisconsin Evangelical Lutheran Synod. The circumstances of her departure are not entirely clear. A possible reason was this denomination's history of anti-Catholic sentiment. For instance, her former church's website stated that "Scripture . . . reveals that the Papacy is the Antichrist."[31]

In the scramble for people's allegiance (and votes) that marks the American political process, secularists are blessed with a gift: an opponent that has a hard time not alienating the three-quarters of Americans who are not like them. Moreover, Revivalists tend to do this when the cameras are rolling. A clever secular leadership would use this propensity to its advantage.

One of the most powerful weapons secularism wields is *the im-*

moderation of its opponents. There is a way – and it may take a lot of patience and cunning – for clever secularists to quietly empower different traditionalist religious groups to do what they have always done: bicker among themselves and outrage moderates so much that the importance of a secular state becomes clear to all.

10. Grin and Bear It: The leadership of the secular movement will have to master an admittedly frustrating tactic: knowing when not to complain or make noise about an issue even though such a response appears justified or, at the very least, would feel good.

In the previous pages we reviewed many examples of how secularists, often unwittingly, assisted their adversaries. As we noted in the case of John McCain's 2008 electoral resurrection, Revivalists rallied around the anti-secular flag. The same logic applied when evangelicals used secular humanism as a convenient boogeyman in the 1980s and when the constitutionality of the Pledge of Allegiance was questioned. The very term *secular* irrationally engages and enrages some in this country; secular leadership needs to be cognizant of that.

This, obviously, doesn't mean that secular pressure groups should avoid activities that might rouse Revivalists. It does mean that from a weak starting position they have to pick and choose their fights very carefully. For example, this book has called for a drawing down of crèche activism. Other indignities that secularists just might have to grin and bear include the irritating public "Christing-up" activities of politicians.

Yes, it was ill-advised of President Obama to say publicly on Easter Sunday, "We're reminded that in that moment, he took on the sins of the world – past, present, and future – and he extended to us that unfathomable gift of grace and salvation through his death and resurrection."[32] Yes, it was ludicrous for Texas governor Rick Perry in August 2011 to invite his fellow governors to join him in a day of Jesus-centered prayer and fasting for "a nation in crisis."[33]

But as infuriating and irresponsible as such activities are, they are not illegal. Public servants have a right to express religious sentiments in public. The truth is, not a whole lot can be done about that. There is no constitutional sanction against Obama or Perry, as private citizens, doing God talk. Interestingly, the Freedom from Religion Foundation tried to prevent Obama from authorizing a national day of prayer. The Federal Court of Appeals for the Seventh Circuit in Wisconsin snippily dismissed the foundation's case, arguing that "hurt feelings differ from legal injury."[34]

Nothing in the Constitution bars a church from engaging in public policy advocacy, and churches have done so across American history.[35] A scholar points out "the U.S. Constitution does not say that religion must be a wholly private matter, and I see no evidence that most religious citizens ever agreed, even tacitly, to treat religion as if it were."[36] Indeed there are many gray areas in church-state relations. Secularists, for now, need to focus solely on the significant trespasses. To invoke the words of Mr. Madison, secularists must learn to let go of "unessential points."

11. Fight Anti-Atheist Prejudice: Throughout this study we have observed that anti-atheist sentiment in this country is pervasive and real. Eradicating this prejudice must be a major part of the new coalition's educational efforts and outreach.

We do not have the solution to this problem – irrational dislike of a group is never easy to uproot. Chris Stedman's suggestion that atheists create meaningful relationships with people who are not like them is important. Stedman, a fellow of the humanist chaplaincy at Harvard University, draws parallels with the struggle for gay rights. "The LGBTQ community," he notes, "has learned that engaged relationships change people's hearts and minds, and this is a model that can be applied to the issue of anti-atheist bias as well."[37]

In addition to the interpersonal challenge there is a legal one as well. Our analysis indicated that many of the judicial difficulties confronting atheists and agnostics have to do with the fact that the Constitution does not recognize them. It tacitly views religion as a good and assumes that its audience is composed entirely of believers who want to get to God. When Justice Rehnquist remarked that the Establishment Clause "did not require government neutrality between religion and irreligion," he cut to the heart of the matter: the foundational documents do not directly acknowledge nonbelief as a component of a legitimate identity.[38] American secularism needs to discover a legal grounding for ensuring the equal rights of the nation's atheist citizens.

For what it's worth, the model may be grafted from recent European attempts to stipulate fundamental freedoms. For example, the Concluding Document of 1989, produced in Vienna and devoted to security and cooperation in Europe, stated, "In order to ensure the freedom of the individual to profess and practise religion or belief, the participating States will, *inter alia,* take effective measures to prevent and eliminate discrimination against individuals or communities on the grounds of religion or belief in the recognition, exercise, and enjoyment of human rights and fundamental freedoms in all fields of civil, political, economic, social, and cultural life, and to ensure *the effective equality between believers and non-believers.*"[39] Equality between believers and nonbelievers is essential to a movement that equally values freedom of, and from, religion.

12. More Ideas, More History, More Clarity: For reasons that are not entirely clear, secularism is currently suffering from a massive brain drain. There exists, for example, very little substantive scholarship on secularism. Scads of social scientists are at work on what is known as the "secularization hypothesis"—the theory as to how the role of religion in public life has diminished across history. Elsewhere, veritable battalions of postmodern and postcolonial critics of secularism lampoon it as "an authoritative discourse" (a term of derision, and even shame, in such quarters) that can be described as "ethnocidal" and "ethnophobic," "hegemonic," patriarchal, Islamophobic, Eurocentric, and so forth.[40]

But serious studies of a subject like the growth and development of American secularism could be counted on one hand. In the previous chapters we leaned on three very important works: Philip Hamburger's *Separation of Church and State,* Noah Feldman's *Divided by God,* and Martha Nussbaum's *Liberty of Conscience.* If it weren't for these fine investigations, there would be almost no scholarly, longitudinal, big-picture analyses of our subject.

Also, secular intellectuals seem to be behind in dealing with emerging theoretical flashpoints. Because of the rise of the Tea Party and Revivalism, we will be hearing a lot in the next decade about states' rights, constitutional originalism, and accommodationism. In all of these ideological growth sectors, new secular ideas and strategies are desperately needed.

The same holds true for communalism, or the tendency of religious citizens to enter the religious sphere as a community. This, as we have seen, is secularism's kryptonite, especially when election day rolls around. How to deal with faith communities that reject the idea that the individual is the fundamental unit of democracy is a more daunting problem than a crèche occupying public space. "The secular state," writes one commentator, "views the individual as a citizen, and not as a member of a particular religious group";[41] it also has the damnedest time figuring out how to deal with groups who reject that logic. Here again, bold and original ideas are needed to solve an urgent problem.

Of course, these twelve suggestions could be multiplied twelve times over without exhausting the problems secularists need to address. The challenges ahead are daunting; one hopes that secularists of the future exhibit more ingenuity, energy, and awareness than the generation that preceded them.

The End of Plenitude

In the middle decades of the twentieth century American secularism accelerated, took off, soared, and boomed sonically. Its momentous Concorde-like flight occurred during a postwar era characterized by confidence, introspection, and plenitude.

Those qualities powered some stunning gains for secularism. For the first time in its history, the United States could try to live up to its high constitutional standards, especially the clause that spoke of prohibiting an establishment of religion. The Warren Court would no longer permit the country to be, either officially or unofficially, a Christian (meaning Protestant) nation. Its holdings were both a cause and an effect of unprecedented sensitivity to the rights of religious minorities. This could be seen in everything from the election of a Catholic president, to a gradual decrease in anti-Semitism, to the expansion of the nation's religious marketplace to include Buddhists, Hare Krishnas, and non-monotheistic "others."

Secularish people in the 1960s, 1970s, and 1980s could not be faulted for thinking that the worst was behind them. The long-enduring miasma of religious wars and persecution seemed to be blowing out to sea. The era of Protestant establishment – a comparatively untreacherous one, as far as such things go – was drawing down. The nation's more conservative denominations seemed to be playing by the secular "rules of the game." Even a figure like Reverend Falwell *voluntarily* stayed clear of the public square.

And then he and countless others, here and abroad, had a change of heart, or saw an opening, or were riled into action. Whatever the case, when the global Revival made landfall at the end of the 1970s, secularists were caught completely off guard.

Everything about Revivalism befuddled them. The throngs chanting "*Marg bah shah*" and burning flags in front of the U.S. Embassy during the Iranian hostage crisis were utterly incomprehensible. What were those guys so worked up over? Militant settler movements in Israel were puzzling. Those Jews were nothing like the placid and moderate class of professional secular Jews in America. Stateside, the televangelists and their megachurches were so theatrically over-the-top that they appeared to be some type of ingenious spoof.

But that was just the shock-and-awe component of the program. In the United States the Revivalists dismantled a foundational Protestant doctrine that had been in existence for centuries. Secularism as we know it began in the Reformation when a sin-obsessed Luther outsourced the problem of maintaining order to the prince. Over the past few decades, however, Christian Revivalists in this country have undertaken to alleviate sinfulness by becoming the prince themselves. Their crusade to *be* caesar has succeeded beyond anyone's wildest dreams, perhaps even their own.

The age of confidence, introspection, and plenitude is over. We may be its children and grandchildren, but that age is over. What lies ahead is not a future of boldly expanding the vision of the architects of secularism. Rather, the sober task at hand is one of consolidation – a firm and dignified defense of the imperiled secularish virtues of moderation, toleration, and self-criticism as well as the political conditions that make those virtues possible.

Acknowledgments

The author of this book wishes to thank his two research assistants, Aurora Nou and Sam Harbout, both now accelerating toward what will surely be promising academic careers. Sam Dinger, Hope Ellis, Ben Wormald, and Gwen Schwartz also provided valuable assistance.

The project came to fruition only because of the insight, savvy, and moral support of my agent, William Lippincott. Jenna Johnson of Houghton Mifflin Harcourt is to be lauded for inspired intellectual vision and editing.

My Georgetown colleagues, friends, family, wife, and two children know who they are, what they mean to me, and what they endured over the crazed eighteen months (!) when the ideas I had been working on for a decade were put to paper.

I dedicate this book to my late father-in-law, Pasquale Spadavecchia. He would have read it cover to cover and argued with me about everything (even though he would have probably agreed with most of it).

Jacques Berlinerblau
Washington, DC
January 29, 2012

Notes

Preface

1. Laurie Goodstein, “Omitting Clergy at 9/11 Ceremony Prompts Protest,” New York Times (8 September 2011), http://www.nytimes.com/2011/09/09/nyregion/omitting-clergy-from-911-ceremony-prompts-protest.html.
2. Ibid.
3. David Gibson, “Bloomberg, Faith Groups Face Off over 9/11 Prayers,” Christian Century (6 September 2011), http://www.christiancentury.org/article/2011-09/bloomberg-faith-groups-face-over-911-prayers.
4. Ibid.
5. Ibid.
6. Jacques Berlinerblau, “Mayor Bloomberg’s 9/11 No-Clergy Stand,” Brainstorm blog, Chronicle of Higher Education (12 September 2011), http://chronicle.com/blogs/brainstorm/mayor-bloomberg%E2%80%99s-911-stand-victory-for-american-secularism-or-a-one-off/39157.

   The term “outrage machine” was coined by Rob Boston, “Memo to the Religious Right: On Sept. 11, 2001, Americans Can Pray–Even Without Government Direction,” Wall of Separation blog (8 September 2011), http:/blog.au.org/blogs/wall-of-separation/memo-to-the-religious-right-on-sept-11-2011-americans-can-pray–even.
7. Ibid.
8. Ibid.
9. “Transcript of Obama’s 9/11 Speech,” United Press International (11 September 2011), http://www.upi.com/Top_News/US/2011/09/11/Transcript-of-Obamas-911-speech/UPI-16831315785864/.
10. “10 Years Later, NYC Honors 9/11 Victims with Tribute in Light, Sol-

emn Ceremony," CBS New York (11 September 2011), http://newyork.cbslocal.com/2011/09/11/new-york-city-marks-10th-anniversary-of-911-in-solemn-ceremony/.

11. Eyder Peralta and Mark Memmott, "Ten Years Later, the Nation Remembers the Sept. 11 Attacks," National Public Radio (11 September 2011), http://www.npr.org/blogs/thetwo-way/2011/09/11/140373696/live-blog-10-years-later-the-nation-remembers.

    For the text of the Bixby letter, see "29. Bixby Letter of Condolence: Lincoln to Mrs. Bixby, November 21, 1864," in "Abraham Lincoln, from His Own Words and Contemporary Accounts," National Park Service (2003), http://www.nps.gov/history/history/online_books/source/sb2/sb2x.htm.

12. "9/11: The Tenth Anniversary—Part 6," CBS New York (11 September 2011), http://newyork.cbslocal.com/video/6241792-911-the-tenth-anniversary-part-6/.

Introduction: Is Secularism Dead?

Epigraph: Caspar Melville, "Mix and Match Secularism" (2 July 2011), guardian.co.uk.

1. The premodern, even biblical roots of secularism and the related but entirely distinct concept of secularization have been studied at length. A particularly fine primer is the excellent work of T. N. Madan, *Modern Myths, Locked Minds: Secularism and Fundamentalism in India* (Delhi: Oxford University Press, 1997), particularly his concise history of premodern secular impulses (pp. 5–25). Major works exploring these premodern taproots, albeit through the lens of secularization, include Harvey Cox's *The Secular City: Secularization and Urbanization in Theological Perspective* (New York: Macmillan, 1965) and Peter Berger's *The Sacred Canopy: Elements of a Sociological Theory of Religion* (New York: Anchor, 1969).
2. Cathy Lynn Grossman, "Pope Tells Bishops to Fight Secular Ideology," *USA Today* (16 April 2008), http://www.usatoday.com/news/religion/2008-04-16-popeside_N.htm. The pope has written quite interestingly on secularism—see Joseph Ratzinger, *Europe Today and Tomorrow: Addressing the Fundamental Issues*, translated by Michael

Miller (San Francisco: Ignatius Press, 2005), p. 99, wherein he defines the secular state as one that tries to operate by "reason alone."

3. He develops this line of analysis in Joseph Ratzinger and Marcello Pera, *Without Roots: The West, Relativism, Christianity, Islam*, translated by Michael Moore (New York: Basic Books, 2007), p. 80. For an Orthodox variant on this topic, see Hilarion Alfeyev, "European Christianity and the Challenge of Militant Secularism," *Ecumenical Review* 57 (January 2005), pp. 82–91.
4. See, for example, Guenter Salter, "The Sordid Results of Humanism," in *The Humanist Threat* (Greenville, SC: Unusual Publications, 1987), p. 44.
5. The quote is from Dr. L. Nelson Bell, father-in-law to Billy Graham, in *While Men Slept* (Garden City, NY: Doubleday, 1979), p. 19. The writer here was speaking about "secular humanism." As we shall see in Chapter 11, secular humanism and secularism are synonyms in conservative evangelical discourse. Clerical hatred of secularism, incidentally, is certainly not unprecedented. See Horace Kallen's discussion of midcentury attitudes in *What I Believe and Why–Maybe: Essays for the Modern World*, translated by Alfred Marrow (New York: Horizon Press, 1971), pp. 154–64.
6. Blair's remarks indicate that he is equating secularism with New Atheism. Ruth Gledhill, "Blair Discusses Threat to World Religions," *The Guardian* (7 October 2009), http://www.timesonline.co.uk/tol/comment/faith/article6864775.ece.
7. For Mitt Romney's "Faith in America" speech, delivered on December 6, 2007, see http://www.npr.org/templates/story/story.php?storyId=16969460. Romney did, however, indicate that he viewed radical jihadism as "infinitely worse."
8. Dana Milbank, "Washington Sketch: Gingrich Is No Longer Conservative About Showing His Religion," *Washington Post* (10 November 2009), http://www.washingtonpost.com/wp-dyn/content/article/2009/11/09AR2009110903302.html.
9. Newt Gingrich, *To Save America: Stopping Obama's Secular-Socialist Machine* (Washington, DC: Regnery, 2011).
10. Barack Obama, *The Audacity of Hope: Thoughts on Reclaiming the American Dream* (New York: Crown Publishers, 2006), p. 39.
11. Ibid.

12. Herbert London, *America's Secular Challenge: The Rise of a New National Religion* (New York: Encounter Books, 2008), p. 2.
13. See, for example, his remarks at the National Cathedral, in "A New Century: A New Reformation," *Washington National Cathedral* (27 January 2008), http://www.nationalcathedral.org/learn/forumTexts/SF080127T.shtml.
14. Susan Brooks Thistlethwaite, *Dreaming of Eden: American Religion and Politics in a Wired World* (New York: Palgrave Macmillan, 2010), p. 58.
15. Ashis Nandy, "Closing the Debate on Secularism: A Personal Statement," in *The Crisis of Secularism in India,* edited by Anuradha Dingwaney Needham and Rajeswari Sunder Rajan (Durham, NC: Duke University Press, 2007), p. 112. Nandy also makes the point that "the true heroes of secularism in the last hundred years have been Adolf Hitler, Joseph Stalin, Mao Tse-tung, and Pol Pot" (p. 111). Drawing the link between secularism and totalitarianism is Emmet Kennedy, *Secularism and Its Opponents, from Augustine to Solzhenitsyn* (New York: Palgrave, 2006), pp. 8–9.
16. Daniel Bell, "The Return of the Sacred: The Argument About the Future of Religion," *Bulletin of the American Academy of Arts and Sciences,* vol. 31, no. 6 (March 1978), pp. 29–55. Robert Wuthnow, *Rediscovering the Sacred: Perspectives on Religion in Contemporary Society* (Grand Rapids, MI: Eerdmans, 1992).
17. Peter Berger, "Secularism in Retreat," in *Islam and Secularism in the Middle East,* edited by John Esposito and Azzam Tamimi (New York: New York University Press, 2000), p. 41.
18. John Micklethwait and Adrian Wooldridge, *God Is Back: How the Global Revival of Faith Is Changing the World* (New York: Penguin Press, 2009).
19. Eric Kaufmann, *Shall the Religious Inherit the Earth?: Demography and Politics in the Twenty-first Century* (London: Profile Books, 2010). Also see Eric Kaufmann, "Demographic Radicalization?: The Religiosity-Fertility Nexus and Politics," *International Studies Association Conference,* 2009, http://www.sneps.net/RD/uploads/1-1-Demographic%20radicalization.pdf.
20. Kaufmann, *Shall the Religious Inherit the Earth?*, p. 130. Also see Caspar Melville, "Battle of the Babies," *New Humanist* 125 (April 2010), http://newhumanist.org.uk/2267/battle-of-the-babies.

21. Kaufmann, "Demographic Radicalization?," p. 14. As a result, the Haredi (the ultra-orthodox) school system in Israel is growing at a rate "39 times greater than that of the state secular schools." Aluf Benn, "Israel's Real Existential Threat Is Arab and Haredi Isolation," *Ha'aretz* (17 February 2010), http://www.haaretz.com/print-edition/opinion/israel-s-real-existential-threat-is-arab-and-haredi-isolation-1.263467.
22. "'In largely Mormon Utah, there are 90 children for every 1,000 women of child-bearing age, compared to only 49 in the socially liberal Vermont of Howard Dean.'" Michael Lind, "Red-State Sneer," *Prospect* (16 December 2004).
23. These events are discussed in detail in Jacques Berlinerblau, *Thumpin' It: The Use and Abuse of the Bible in Today's Presidential Politics* (Louisville, KY: Westminster John Knox Press, 2008). Also see John Green et al. (eds.), *The Values Campaign?: The Christian Right and the 2004 Elections* (Washington, DC: Georgetown University Press, 2006).
24. See Berlinerblau, *Thumpin' It*, p. 12.
25. Cited in William Goodman Jr. and James Price, *Jerry Falwell: An Unauthorized Profile* (Lynchburg, VA: Paris and Associates, 1981), p. 33.
26. Here invoking James Madison's words in "Memorial and Remonstrance Against Religious Assessments," in Forrest Church (ed.), *The Separation of Church and State: Writings on a Fundamental Freedom by America's Founders* (Boston: Beacon Press, 2004), p. 64.
27. Daniel Dennett, "The Bright Stuff," *New York Times* (12 July 2003), p. 11. On how Dennett came up with that dubious figure, see Berlinerblau, *Thumpin' It*, p. 173, n72. Also see Paul Starobin, "The Godless Rise as a Political Force," *National Journal* (7 March 2009), http://www.nationaljournal.com/membera/magazine/the-godless-rise-as-a-political-force-20090307. For a fascinating account of the instability and complexity of the term *atheism*—nuances that generally evade the New Atheists—see Stephen Bullivant, "Research Note: Sociology and the Study of Atheism," *Journal of Contemporary Religion* 23 (2008), pp. 363–68.
28. Jeffrey Stout, "2007 Presidential Address: The Folly of Secularism," *Journal of the American Academy of Religion* 76 (2008), p. 540.
29. For some basic overviews of secularization, see Karel Dobbelaere, "Secularization," in *Encyclopedia of Religion and Society*, edited by Wil-

liam Swatos Jr. (Walnut Creek, CA: Altamira Press, 1998), pp. 452–56; Lindsay Jones (ed.), "Secularization," in *The Encyclopedia of Religion,* 2nd ed., vol. 12 (New York: Thomson Gale, 2005), pp. 8214–219; Frank Lechner, "Secularization," in *The Encyclopedia of Protestantism,* vol. 4, edited by Hans Hillerbrand (New York: Routledge, 2004), pp. 1701–707; D. Howard Smith, "Secularization," in *A Dictionary of Comparative Religion,* edited by S.G.F. Brandon (London: Weidenfeld & Nicolson, 1970), pp. 568–69; Bryan Wilson, "Secularization," in *The Encyclopedia of Religion,* vol. 13, edited by Mircea Eliade (New York: Macmillan Publishing Company, 1987), pp. 159–65; Davina Allan, "Secularization," in *Key Ideas in Human Thought,* edited by Kenneth McLeish (New York: Facts on File, 1993), p. 668. Also see, for example, Christian Smith (ed.), *The Secular Revolution: Power, Interests, and Conflict in the Secularization of American Public Life* (Berkeley: University of California Press, 2003).

30. Cited in Heiko Oberman, *Luther: Between Man, God, and the Devil* (New Haven, CT: Yale University Press, 2006), p. 105.
31. Martha Nussbaum, *Liberty of Conscience: In Defense of America's Tradition of Religious Equality* (New York: Basic Books, 2008), p. 69.
32. A point made early and often by John Garrett in *Roger Williams: Witness Beyond Christendom, 1603–1683* (New York: Macmillan, 1970).
33. Roger Williams, *The Bloudy Tenent of Persecution for Cause of Conscience Discussed and Mr. Cotton's Letter Examined and Answered* (LaVergne, TN.: Kessinger Publishing, 2010), p. 49.
34. On this important point see Garrett, *Roger Williams.*
35. It is likely that Locke had read Williams. Nussbaum, *Liberty of Conscience,* p. 67.
36. For example, Locke could offer "support for the establishment of a national Church of England, to be headed by, and partly funded by, the Crown." See David McCabe, "John Locke and the Argument Against Strict Separation," *The Review of Politics* 59 (1997), p. 248.
37. Owen Chadwick, *The Secularization of the European Mind in the Nineteenth Century* (Cambridge: Cambridge University Press, 1975), p. 25. Chadwick's work, incidentally, is one of the few classics on the subject of secularism.
38. Bernhard Fabian, "Jefferson's Notes on Virginia: The Genesis of Query xvii, 'The Different Religions Received into that State?,'" *William and*

Mary Quarterly 12 (1995), p. 138. Jack Rakove, "Beyond Locke, Beyond Belief: The Nexus of Free Exercise and Separation of Church and State," in *Religion, State, and Society*, edited by Robert Fatton Jr. and R. K. Ramazani (New York: Palgrave Macmillan, 2009), pp. 37–52.

39. Saint Augustine, *Concerning the City of God Against the Pagans*, translated by Henry Bettenson (New York: Penguin, 2003), book I, p. 5.

1. What Is Secularism? (The Basic Package)

Epigraph: T. N. Madan, in *Secularism and Its Critics*, edited by Rajeev Bhargava (Delhi: Oxford University Press, 2005).

1. John Keane, "The Limits of Secularism," in Esposito and Tamimi, *Islam and Secularism in the Middle East*, pp. 29, 34.
2. First Amendment Center, "'07 Survey Shows Americans' Views Mixed on Basic Freedoms," *Firstamendmentcenter.org* (24 September 2007), http://www.firstamendmentcenter.org/07-survey-shows-americans-views-mixed-on-basic-freedoms.
3. The Pew Forum on Religion and Public Life, "Lift Every Voice: A Report on Religion in American Public Life, 2002" (Washington, DC: The Pew Forum, 2001), p. 36.
4. Linda Lyons, "Americans Indivisible on Pledge of Allegiance," *Gallup* (4 May 2004), http://www.gallup.com/poll/11551/Americans-Indivisible-Pledge-Allegiance.aspx.
5. "Most Americans Prefer 'Merry Christmas' to 'Happy Holidays,'" *Rasmussen Reports* (29 November 2009), http://www.rasmussenreports.com/public_content/lifestyle/holidays/november_2009/most_americans_prefer_merry_christmas_to_happy_holidays. Studies also indicate that Americans may be growing wary of involving religion in politics. Pew Forum on Religion and Public Life, "More Americans Question Religion's Role in Politics," *Pew Research Center* (21 August 2008), http://pewforum.org/Politics-and-Elections/More-Americans-Question-Religions-Role-in-Politics.aspx.
6. On the fealty of the French to their secular system, see Nathalie Caron, "*Laïcité* and Secular Attitudes in France," in *Secularism and Secularity: Contemporary International Perspectives*, edited by Barry Kosmin

and Ariela Keysar (Hartford, CT: Institute for the Study of Secularism in Society and Culture, 2007), pp. 113–24.

7. The Syrian Constitution mandates separation of church and state. Scott Merriman, "Syria," in *Religion and the State: An International Analysis of Roles and Relationships* (Santa Barbara, CA: ABC CLIO, 2009), pp. 307–8; Catherine Field, "Religions Thrive in a Troubled Land," *New York Times* (27 April 2011), http://www.nytimes.com/2011/04/28/opinion/28iht-edfield28.html?scp=1&sq=secularism%20in%20syria&st=cse.
8. For example, see Stanley Tambiah, "The Crisis of Secularism in India," in Bhargava, *Secularism and Its Critics*, pp. 418–53, as well as many of the other fine articles in this landmark volume.
9. Neil MacFarquhar, "Radicals' Turn to Democracy Alarms Egypt," *New York Times* (1 April 2001), p. A1.
10. See the interesting analysis of Fadi Hakura, "What Can Rescue the Arab Spring?" *Christian Science Monitor* (11 May 2011), http://www.csmonitor.com/Commentary/Opinion/2011/0510/What-can-rescue-the-Arab-Spring.
11. Martin Luther understood this need for balance. He knew that if a secular government "overreached itself," the results would be "unbearable and horrifying." The same held true for the reverse approach. Martin Luther, "On Secular Authority," in *Luther and Calvin on Secular Authority*, edited by Harro Höpfl (Cambridge: Cambridge University Press, 2006), p. 22.
12. Ibid.
13. James Madison, "To F. L. Schaeffer," in *Letters and Other Writings of James Madison, Fourth President of the United States, in Four Volumes. Vol. 3: 1816–1828* (New York: R. Worthington, 1884), pp. 242–43.
14. Luther, "On Secular Authority," in Höpfl, *Luther and Calvin on Secular Authority*, p. 11.
15. Ibid., p. 10.
16. Martin Luther, "Psalm 82," in Jaroslav Pelikan (ed.), *Luther's Works. Vol. 13: Selected Psalms II* (Saint Louis, MO: Concordia Publishing House, 1956), pp. 44–45.
17. John Calvin, "On Civil Government," in Höpfl, *Luther and Calvin on Secular Authority*, p. 49. Also see John Witte Jr., "That Serpentine

Wall of Separation," *Michigan Law Review* 101 (2003), p. 1884 (1869–1905).

18. See Matthew 22:1, Mark 12:17, and Luke 20:25 in the King James Version.
19. New Revised Standard Version (NRSV).
20. Also see Hebrews 13:17. The same mindset compels the author of 1 Timothy 2:1 to suggest that prayers be made "for kings and all who are in high positions, so that we may lead a quiet and peaceable life in all godliness and dignity."
21. Martin Luther, *Luther's Works. Vol. 25: Lectures on Romans, Glosses, and Scholia,* edited by Hilton Oswald (Saint Louis, MO: Concordia, 1955), p. 109.
22. Jean-Jacques Rousseau, *The Social Contract,* translated by Maurice Cranston (New York: Penguin, 1983), p. 184.
23. Evidence indicates that Luther's practice did not align with his theory. In his dealings with groups considered heretical, such as the Anabaptists, he had no problem in urging the state to punish the dissenters. He viewed Anabaptists as engaging in sedition, thus blurring the line between church and state. On this see John Oyer, *Lutheran Reformers Against Anabaptists: Luther, Melanchton, and Menius and the Anabaptists of Central Germany* (The Hague: Martinus Nijhoff, 1964), p. 136. So even though he claimed that "the use of force can never prevent heresy," he urged the full coercive weight of the state to be levied against the "robbing and murdering hordes of peasants." Luther, "On Secular Authority," in Höpfl, *Luther and Calvin on Secular Authority,* p. 30; Martin Luther, "Against the Robbing and Murdering Hordes of Peasants," in *Documents from the History of Lutheranism, 1517–1750,* edited by Eric Lund (Minneapolis: Fortress Press, 2002), p. 44.

    It gets worse. Lutheran rulers in the sixteenth century "transferred to the secular authorities jurisdiction over matters that previously were governed by the spiritual law of the Church of Rome." Harold Berman, "The Spiritualization of Secular Law: The Impact of the Lutheran Reformation," *Journal of Law and Religion* 14 (1999–2000), p. 317.
24. Martin Luther, "To the Christian Nobility of the German Nation Concerning the Reform of the Christian Estate," in Garrett Ward Sheldon (ed.), *Religion and Politics: Major Thinkers on the Relation of Church and State* (New York: Peter Lang, 1990), p. 71.

25. Luther, "On Secular Authority," in Höpfl, *Luther and Calvin on Secular Authority*, p. 23. For Locke's similar surmise, see *A Letter Concerning Toleration* (Indianapolis: Hackett Publishing Company, 1983), p. 26.
26. It must be stressed, however, that all of our architects, in ways that are sometimes astonishing and embarrassing, did not practice what they theorized. For example, Locke could offer "support for the establishment of a national Church of England, to be headed by, and partly funded by, the Crown." See David McCabe, "John Locke and the Argument Against Strict Separation," *Review of Politics* 59 (1997), p. 248.
27. Perez Zagorin, *How the Idea of Religious Toleration Came to the West* (Princeton, NJ: Princeton University Press, 2003), p. 254.
28. For one example, see Michael Feldberg, *The Philadelphia Riots of 1844: A Study of Ethnic Conflict* (Westport, CT: Greenwood Press, 1975), p. 85.
29. Derek Davis, "Religion, Regulation of," in *Encyclopedia of Religion in America. Vol. 4: Q–Z,* edited by Charles Lippy and Peter Williams (Washington, DC: CQ Press, 2010), p. 1849.
30. R. J. Knecht, *The French Wars of Religion, 1559–1598,* 2nd ed. (New York: Longman, 1996), pp. 42–51. Barbara Diefendorf, *The Saint Bartholomew's Massacre: A Brief History with Documents* (Boston: Bedford/St. Martin's, 2009), p. 102.
31. C. V. Wedgwood, *The Thirty Years War* (New York: New York Review Books, 2005), p. 496. Also see Geoff Mortimer, *Eyewitness Accounts of the Thirty Years War, 1618–1648* (New York: Palgrave, 2002). On the role of the Treaty of Westphalia in the secular state system, see Daniel Philpott, "The Challenges of September 11 to Secularism in International Relations," *World Politics* 55 (2002), pp. 66–95.
32. See the statistics cited in Charles Carlton, *Going to the Wars: The Experience of the British Civil Wars, 1638–1651* (London: Routledge, 1992), p. 214.
33. Thomas Hamm, "Quakers: Through the Nineteenth Century," in Lippy and Williams, *Encyclopedia of Religion in America,* vol. 4, pp. 1807–814. Also see T. Jeremy Gunn, "Religious Freedom and *Laïcité:* A Comparison of the United States and France," *Brigham Young University Law Review* (Summer 2004), p. 443.
34. Roger Williams, of course, did not live through all of the events listed here. The most enlightening discussion of the Salem catastrophe re-

mains that of Paul Boyer and Stephen Nissenbaum, *Salem Possessed: The Social Origins of Witchcraft* (Cambridge, MA: Harvard University Press, 1974).

35. These remarks are courtesy of the Unitarian minister Richard Price in 1785, "Of Liberty of Conscience and Civil Establishment of Religion," in Church, *The Separation of Church and State,* pp. 147, 151.
36. James Madison, "Memorial and Remonstrance Against Religious Assessments," in Church, *The Separation of Church and State,* p. 65.
37. Cited in Edwin Gaustad, *Liberty of Conscience: Roger Williams in America* (Grand Rapids, MI: Eerdmans, 1991), p. 194.
38. Cited in Melvin Urofsky, *Religious Freedom: Rights and Liberties Under the Law* (Santa Barbara, CA: ABC-CLIO, 2002), p. 118.
39. "Constitution of 1791," in Frank Maloy Anderson, *The Constitutions and Other Select Documents Illustrative of the History of France, 1789–1901* (Minneapolis: H. W. Wilson, 1904), pp. 59–60. The 1789 text was incorporated into the constitution of 1791.
40. "Decree on the Separation of Church and State of January 23, 1918," in Richard Marshall Jr. (ed.), *Aspects of Religion in the Soviet Union, 1917–1967* (Chicago: University of Chicago Press, 1971), pp. 437–38.
41. United Nations General Assembly, International Covenant on Civil and Political Rights (16 December 1966), Article 18, http://www2.ohchr.org/english/law/ccpr.htm.
42. As one critic poses it, secularism is "non-religious rather than anti-religious." Andre Beteille, "Secularism and Intellectuals," *Economic and Political Weekly* 29 (1994), p. 566.
43. Max Weber, *Economy and Society: An Outline of Interpretive Sociology,* edited by Guenther Roth and Claus Wittich (Berkeley: University of California Press, 1978), p. 1162.
44. Locke, *A Letter Concerning Toleration,* p. 23.
45. Ratzinger and Pera, *Without Roots,* p. 62.
46. Luther, "On Secular Authority," in Höpfl, *Luther and Calvin on Secular Authority,* p. 5.
47. As James Estes notes, "Luther steadfastly maintained that secular office per se entailed no authority whatever in spiritual matters." *Peace, Order, and the Glory of God: Secular Authority and the Church in the Thought of Luther and Melanchton, 1518–1559* (Leiden: Brill, 2006), p. 13.

48. Williams, *The Bloudy Tenent*, p. 13. For more on Williams, see Nussbaum, *Liberty of Conscience*.
49. Locke, *A Letter Concerning Toleration*, p. 35.
50. Thomas Jefferson, "Virginia Statute for Religious Freedom," in Church, *The Separation of Church and State*, p. 74. As part of our footnote series on contradictions between the thought and the actions of the architects, we observe that Jefferson once rejected the testimony of an atheist in a court of law simply because the witness was an atheist. Sanford Kessler, "Locke's Influence on Jefferson's 'Bill for Establishing Religious Freedom,'" *Journal of Church and State* 25 (1983), p. 239.
51. Madison, "Memorial and Remonstrance," in Church, *The Separation of Church and State*, p. 61. This opinion is seconded in 1660 by Samuel Pufendorf, a theorist whose work bore interesting affinities to those of Locke, who reasoned that it was not "appropriate for a magistrate to apply force in enjoining a religion upon men." Pufendorf, "Observation 1," in *The Political Writings of Samuel Pufendorf*, edited by Craig Carr (New York: Oxford University Press, 1994), p. 74.
52. Some reversed the equation as well.
53. Locke, *A Letter Concerning Toleration*, p. 48.
54. As is well known, Locke would not extend that toleration to atheists. Moreover, his views on Catholics are uncharitable at best. Secular posterity would get around to cleaning up that mess.
55. Ibid., p. 23.
56. Luther, "On Secular Authority," in Höpfl, *Luther and Calvin on Secular Authority*, p. 25.
57. Ibid., p. 24.
58. Williams, *The Bloudy Tenent*, p. 10. Nussbaum, *Liberty of Conscience*, p. 37.
59. Locke, *A Letter Concerning Toleration*, p. 38.
60. Rakove, "Beyond Locke, Beyond Belief," in Fatton and Ramazani, *Religion, State, and Society*, pp. 37–52. "But where he stopped short," exclaimed Jefferson, paying homage to Locke, "we may go on." Fabian, "Jefferson's Notes on Virginia," p. 138.
61. Thomas Jefferson, "Notes on the State of Virginia," in Church, *The Separation of Church and State*, pp. 51–52.
62. Michael McConnell et al., *Religion and the Constitution* (New York: Aspen Publishers, 2002), p. 84.

63. As the legal theorist Michael McConnell points out, "There is no free exercise right to kidnap another person for the purpose of proselytizing, or to trespass on private property – be it an abortion clinic or a defense contracting plant – to protest immoral activity. Conduct on public property must be peaceable and orderly, so that the rights of others are not disturbed." McConnell et al., *Religion and the Constitution,* p. 108. A similar principle was articulated in *Davis v. Beason* in 1890: "However free the exercise of religion may be, it must be subordinate to the criminal laws of this country." Ibid., p. 143.
64. For a detailed discussion of the 1879 *Reynolds v. United States* case, see Noah Feldman, *Divided by God: America's Church-State Problem – and What We Should Do About It* (New York: Farrar, Straus, Giroux, 2005), pp. 99–110.
65. *Employment Div. v. Smith,* 494 U.S. 872, 886 (1990).
66. On the statist dimensions of secularism, see Nandy, "The Politics of Secularism and the Recovery of Religious Tolerance," in Bhargava, *Secularism and Its Critics,* pp. 321–44, esp. p. 333.
67. Locke, *A Letter Concerning Toleration,* pp. 30–32.
68. Ibid., p. 30.
69. Ibid., p. 55.
70. Ibid., p. 26 (emphasis in original).
71. This is one of the most common criticisms of secularism. See Janet Jakobsen and Ann Pellegrini, "Introduction: Times like These," in *Secularisms,* edited by Janet Jakobsen and Ann Pellegrini (Durham, NC: Duke University Press, 2008), p. 13. T. N. Madan, "Secularism in Its Place," in Bhargava, *Secularism and Its Critics,* p. 309.
72. Roger Williams, "The Hireling Ministry None of Christ's," in *On Religious Liberty: Selections from the Work of Roger Williams,* edited by James Davis (Cambridge, MA: Harvard University Press, 2008), p. 257. The rather extreme and unyielding dimensions of Williams's character are nicely sketched in Timothy Hall, *Separating Church and State: Roger Williams on Religious Liberty* (Urbana: University of Illinois Press, 1998).
73. For example, Egypt News, "Egypt Pope Shenouda Calls for Secular State," *Egypt.com* (24 April 2011), http://news.egypt.com/en/2011042414436/news/-egypt-news/egypt-pope-shenouda-calls

-for-a-secular-state.html. This point is also made in Nussbaum, *Liberty of Conscience,* p. 15.

74. A point that could be gleaned by reading Roger Olson's discussion of the restructuring of evangelicalism and "neo-evangelicalism" in the aftermath of the Scopes trial. *The Westminster Handbook to Evangelical Theology* (Louisville, KY: Westminster John Knox Press, 2004), p. 5. Also see John Fea, *Was America Founded as a Christian Nation?: A Historical Introduction* (Louisville, KY: Westminster John Knox Press, 2011), p. 44.
75. For a very similar point about secularism creating the preconditions for its own demise, see the discussion between Caspar Melville and Eric Kaufmann in "Battle of the Babies."

2. *Were the Founders Secular?*

Epigraph: Mariah Blake, "Revisionaries: How a Group of Texas Conservatives Is Rewriting Your Kids' Textbooks," *Washington Monthly* (January–February 2010), http://www.washingtonmonthly.com/features/2010/1001.blake.html.

1. James C. McKinley Jr., "Texas Conservatives Win Curriculum Change," *New York Times* (12 March 2010), http://www.nytimes.com/2010/03/13/education/13texas.html. This attempt to reconstruct the narrative of this controversy is severely hampered by the complexity of the original documents themselves. All references to the Texas State Board of Education documents are included here, though it is often not clear from these documents when certain decisions were made and why.
2. Texas State Board of Education, "Proposed Revisions to 19 TAC Chapter 113, *Texas Essential Knowledge and Skills for Social Studies,* Subchapter C, *High School,* and 19 TAC Chapter 118, *Texas Essential Knowledge and Skills for Economics with Emphasis on the Free Enterprise System and Its Benefits,*" Subchapter A, *High School,* p. 46, http://www.tea.state.tx.us/index2.aspx?id=3643.
3. Ibid., p. 8.
4. Ibid., p. 42.

5. Ibid., p. 25.
6. Ibid., p. 25.
7. Texas Freedom Network, "Blogging the Social Studies Debate IV" (11 March 2010), http://tfninsider.org/2010/03/11/blogging-the-social-studies-debate-iv/.
8. Even when it was assumed that Jefferson would no longer be included in the "Enlightenment section," board members were adamant that he remain in other parts of the curriculum, such as a unit called "Founding Fathers and Patriot Heroes." Texas State Board of Education, "Thomas Jefferson Remains in Social Studies Curriculum" (19 March 2010), http://static.texastribune.org/media/documents/SBOE_statement_Jefferson.pdf.
9. Texas State Board of Education, "Thomas Jefferson Remains in Social Studies Curriculum," *North Texas e-News* (20 March 2010), http://www.ntxe-news.com/cgi-bin/artman/exec/view.cgi?archive=37&num=60884.
10. Russell Shorto, "How Christian Were the Founders?" *New York Times Magazine* (11 February 2010), http://www.nytimes.com/2010/02/14/magazine/14texbooks-t.html?pagewanted=all.
11. James C. McKinley Jr., "Texas Conservatives Seek Deeper Stamp on Texts," *New York Times* (10 March 2010), http://www.nytimes.com/2010/03/11/us/politics/11texas.html. See also Steven Schafersman, "Social Studies Standards Under Attack by State Board of Education Members," *Texas Observer* (10 March 2010), http://www.texasobserver.org/oped/social-studies-standards-under-attack-by-state-board-of-education-members.
12. Robert Mackey, "Textbooks a Texas Dentist Could Love," *The Lede* blog, *New York Times* (12 March 2010), http://thelede.blogs.nytimes.com/2010/03/12/textbooks-a-texas-dentist-could-love/.
13. McKinley, "Texas Conservatives Win Curriculum Change."
14. Cited in William Lee Miller, *The First Liberty: America's Foundation in Religious Freedom* (Washington, DC: Georgetown University Press, 2003), p. 98.
15. Alf Mapp Jr., *The Faith of Our Fathers: What America's Founders Really Believed* (Lanham, MD: Rowman and Littlefield, 2003), p. 108.
16. On Hamilton, see Gregg Frazer, "Alexander Hamilton, Theistic Rationalist," in *The Forgotten Founders on Religion and Public Life,* edited

by Daniel L. Dreisbach, Mark D. Hall, and Jeffry H. Morrison (Notre Dame, IN: University of Notre Dame Press, 2009), pp. 101–24.

17. *The Declaration of Independence and the Constitution of the United States of America* (Washington, DC: National Defense University, 1995), p. 77.
18. Elizabeth Tenety, "Separation of Church and State Questioned by Christine O'Donnell," *Washington Post* (19 October 2010), http://voices.washingtonpost.com/44/2010/10/separation-of-church-and-state.html.
19. Thomas Jefferson, "A Wall of Separation," in Church, *The Separation of Church and State,* p. 130.
20. Ibid.
21. Roger Williams, *Mr. Cotton's Letter, Lately Printed, Examined, and Answered* (London: n.p., 1644; eebo.chadwyck.com), p. 45. Also see Mark DeWolfe Howe, *The Garden and the Wilderness: Religion and Government in American Constitutional History* (Chicago: University of Chicago Press, 1965).
22. Daniel Dreisbach, "'Sowing Useful Truths and Principles': The Danbury Baptists, Thomas Jefferson, and the 'Wall' of Separation," *Journal of Church and State* 39 (1997), pp. 484–87. Witte, "That Serpentine Wall of Separation," p. 1876, suggests that Jefferson may have taken the term from Saint Paul.
23. Jefferson, "A Wall of Separation," in Church, *The Separation of Church and State,* p. 130.
24. Tara Ross and Joseph C. Smith Jr., *Under God: George Washington and the Question of Church and State* (Dallas: Spence Publishing, 2008), p. xviii. Daniel Dreisbach, *Thomas Jefferson and the Wall of Separation Between Church and State* (New York: New York University Press, 2002), p. 98.
25. Ibid., pp. 98–99.
26. See the discussion of Donald Drakeman, *Church-State Constitutional Issues: Making Sense of the Establishment Clause* (New York: Greenwood Press, 1991), p. 98.
27. Ross and Smith, *Under God,* p. xviii.
28. Dreisbach, *Thomas Jefferson and the Wall of Separation,* p. 98.
29. Philip Hamburger, *Separation of Church and State* (Cambridge, MA: Harvard University Press, 2002) p. 163. Also see pp. 101, 107, 170, 177–

79. Dreisbach, in *Thomas Jefferson and the Wall of Separation,* p. 30, notes that the letter was reprinted a few days after in "partisan Republican newspapers."
30. Hamburger, *Separation of Church and State,* pp. 163–64. Making a similar point is Dreisbach, *Thomas Jefferson and the Wall of Separation,* p. 51.
31. Hamburger, *Separation of Church and State,* p. 179.
32. Legislative Reference Service, *The Constitution of the United States of America* (Washington, DC: U.S. Government Printing Office, 1964), p. 845.
33. Dreisbach, "'Sowing Useful Truths and Principles,'" p. 459.
34. George Washington, "Farewell Address (Selections)," in Church, *The Separation of Church and State,* p. 119.
35. On Hamilton, see Frazer, "Alexander Hamilton, Theistic Rationalist," in Dreisbach et al., *The Forgotten Founders,* p. 110.
36. Vincent Phillip Muñoz, "Religion and the Common Good: George Washington on Church and State," in *The Founders on God and Government,* edited by Daniel L. Dreisbach, Mark D. Hall, and Jeffry H. Morrison (Lanham, MD: Rowman and Littlefield, 2004), p. 6. A more popular treatment of Washington's views on these issues can be found in Ross and Smith, *Under God.*
37. "Massachusetts Constitution" (extract), in *Church and State in the Modern Age: A Documentary History,* edited by J. F. Maclear (New York: Oxford University Press, 1995), pp. 57–59. Also see John Witte Jr., "One Public Religion, Many Private Religions: John Adams and the 1780 Massachusetts Constitution," in Dreisbach et al., *The Founders on God and Government,* pp. 28, 47. On Adams in this period see also Rosemarie Zagarri, "Mercy Otis Warren on Church and State," in Dreisbach et al., *The Forgotten Founders,* pp. 278–94. And see Zagarri's discussion of how differently the Founder named in her title viewed matters in comparison to Jefferson.
38. "Massachusetts Constitution" (extract), p. 58.
39. John Adams, *The Works of John Adams, Second President of the United States,* vol. 2 (Boston: Charles C. Little and James Brown, 1850), p. 399. Witte, "One Public Religion, Many Private Religions," p. 47. A similar argument about the religious proclivities of the Founders and the relative uniqueness of Jefferson's views can be found in James Hut-

son's articles on "Nursing Fathers" in *Forgotten Features of the Founding: The Recovery of Religious Themes in the Early American Republic* (Lanham, MD: Lexington Books, 2003), pp. 45–71. On the prevalence of this view in New England, see Howe, *The Garden and the Wilderness,* p. 26.

40. John Witte Jr., "The Essential Rights and Liberties of Religion in the American Constitutional Experiment," *Notre Dame Law Review* 71 (1995–1996), p. 379.
41. Barry Alan Shain, "Afterword: Revolutionary-Era Americans: Were They Enlightened or Protestant? Does It Matter?," in Dreisbach et al., *The Founders on God and Government,* p. 277.
42. See William Jay, *The Life of John Jay: With Selections from His Correspondence and Miscellaneous Papers* (New York: J. & J. Harper, 1833), p. 376.
43. "Proposed Revisions to 19 TAC Chapter 113, *Texas Essential Knowledge and Skills for Social Studies,* Subchapter C, *High School,*" color-coded version, published on the Texas Education Agency website for review, p. 3, http://www.tea.state.tx.us/index2.aspx?id=3643. "Muhlenberg, John Peter Gabriel (1746–1807)," *Biographical Directory of the United States Congress,* http://bioguide.congress.gov/scripts/biodisplay.pl?index=M001066.
44. Dreisbach, *Thomas Jefferson and the Wall of Separation,* p. 67.
45. Hutson, *Forgotten Features of the Founding,* p. 167.
46. Dreisbach et al., "Preface," in *The Forgotten Founders,* pp. xiv–xv. Also see Jon Butler, "Why Revolutionary America Wasn't a 'Christian Nation,'" in *Religion and the New Republic: Faith in the Founding of America,* edited by James Hutson (Lanham, MD: Rowman and Littlefield, 2000), p. 188.
47. In Chapter 7 we will look at nativist parties who advocated for separation in the late nineteenth century. Their judicial impact was fairly minimal.
48. Naomi Cohen, *Jews in Christian America: The Pursuit of Religious Equality* (New York: Oxford University Press, 1992), p. 37.
49. Feldberg, *The Philadelphia Riots of 1844,* p. 85.
50. Joseph Story, *Commentaries on the Constitution of the United States,* vol. 3 (Boston: Hilliard and Gray, 1833), sec. 1871, p. 728.
51. Steven Green makes a strong case for the complexities and nuances

in Brewer's thinking on the separation issue. "Justice David Josiah Brewer and the 'Christian Nation' Maxim," *Albany Law Review* 63 (1999–2000), pp. 423–76.

52. David Brewer, *American Citizenship: Yale Lectures* (New York: Charles Scribner's Sons, 1907), pp. 21–22. Davison Douglas, "'Christian Nation' as a Concept in Supreme Court Jurisprudence," in *Religion and American Law: An Encyclopedia,* edited by Paul Finkelman (New York: Garland, 2000), pp. 74–75. On Brewer's perspective, see Hugh Heclo, "Is America a Christian Nation?," in *The Future of Religion in American Politics,* edited by Charles Dunn (Lexington: University Press of Kentucky, 2009), pp. 61–96.
53. Witte, "The Essential Rights and Liberties," p. 389.
54. On the use of the letter in 1853 and 1856, see Dreisbach, "'Sowing Useful Truths and Principles,'" p. 491.
55. Carol Weisbrod, "*Reynolds v. United States,* 98 U.S. (8 Otto) 145 (1879)," in Finkelman, *Religion and American Law,* pp. 417–21.
56. The events of the trial are recounted by Feldman, *Divided by God,* pp. 99–110.
57. Hamburger, *Separation of Church and State,* p. 260.
58. Henry Abraham, "Religion, the Constitution, the Court, and Society: Some Contemporary Reflections on Mandates, Words, Human Beings, and the Art of the Possible," in *How Does the Constitution Protect Religious Freedom?,* edited by Robert Goldwin and Art Kaufman (Washington, DC: American Enterprise Institute for Public Policy Research, 1987), pp. 31–34.
59. James McClellan, "The Making and the Unmaking of the Establishment Clause," in *A Blueprint for Judicial Reform,* edited by Patrick McGuigan and Randall Rader (Washington, DC: Free Congress Research and Education Foundation, 1981), p. 300.
60. *Everson v. Bd. of Educ.,* 330 U.S. 1, 18 (1947).
61. *Lemon v. Kurtzman,* 403 U.S. 602, 627 (1971).
62. The quote cited here was written four decades after *Everson,* but the sentiments still prevail. McClellan, "The Making and the Unmaking of the Establishment Clause," in McGuigan and Rader, *A Blueprint for Judicial Reform,* p. 318. Also see David Ryden, "The Relevance of State Constitutions to Issues of Government and Religion," in *Church-State Issues in America Today. Vol. 1: Religion and Govern-*

ment, edited by Ann Duncan and Steven Jones (Westport, CT: Praeger, 2008), p. 236.

63. Leonard Levy, *The Establishment Clause: Religion and the First Amendment*, 2nd ed. (Chapel Hill: University of North Carolina Press, 1994), p. 225.
64. See, for example, Mark David Hall, "Jeffersonian Walls and Madisonian Lines: The Supreme Court's Use of History in Religion Clause Cases," *Oregon Law Review* 85 (2006), pp. 563–613. Also see Howe, *The Garden and the Wilderness*, p. 11.
65. Texas State Board of Education, "Thomas Jefferson Remains in Social Studies Curriculum" (19 March 2010).
66. Terrence Stutz, "Texas State Board of Education Approves New Curriculum Standards," *Dallas Morning News* (22 May 2010), http://www.dallasnews.com/news/education/headlines/20100521-Texas-State-Board-of-Education-approves-9206.ece. Terrence Stutz, "More Conservative Textbook Curriculum OK'd," *Dallas Morning News* (22 May 2010), http://www.dallasnews.com/news/education/headlines/20100522-More-conservative-textbook-curriculum-OK-d-3498.ece.
67. Texas State Board of Education, §113.44(c)(1)(B), in "Texas Administrative Code (TAC), Title 19, Part 11, Chapter 113. Texas Essential Knowledge and Skills for Social Studies," http://ritter.tea.state.tx.us/rules/tac/chapter113/ch113c.html.
68. Texas State Board of Education, §113.41(c)(8)(B), in "Chapter 113. Texas Essential Knowledge and Skills for Social Studies, Subchapter C. High School," http://ritter.tea.state.tx.us/rules/tac/chapter113/ch113c.html.
69. Stutz, "Texas State Board of Education Approves New Curriculum Standards."

3. Does Secularism Equal Total Separation of Church and State?

Epigraphs: D. E. Smith, "India as a Secular State," in Bhargava, *Secularism and Its Critics*, p. 179. Nicolas Sarkozy, *La république, les religions, l'ésperance: Entretiens avec Thibaud Collin et Philippe Verdin* (Paris: Les Éditions du Cerf, 2004).

1. Making the same point is Frank Guliuzza III, *Over the Wall: Protect-*

ing Religious Expression in the Public Square (Albany: SUNY Press, 2000), p. 52.

2. Veronica Menaldi, "Michael Newdow: An Atheist with a Cause," *Michigan Daily* (14 April 2010), http://www.michigandaily.com/content/after-they-walk-michael-newdow?page=0,0Veronica.
3. "Atheist Appears in Washington Court to Argue Against Inaugural Prayer," *The Guardian* (15 January 2009), http://www.guardian.co.uk/world/2009/jan/15/obama-inauguration-religion.
4. Associated Press, "Atheist Challenges 'In God We Trust,'" *MSNBC.com* (18 November 2005), http://www.msnbc.msn.com/id/10103424/ns/us_news-life/. Also see "Litigant Explains Why He Brought Pledge Suit," *CNN.com* (26 June 2002), http://edition.cnn.com/2002/LAW/06/26/Newdow.cnna/, and Kimberly Winston, "Despite Abysmal Track Record Calif. Atheist Keeps Suing," *Christian Century* (21 December 2010), http://www.christiancentury.org/article/2010-12/despite-abysmal-track-record-calif-atheist-keeps-suing. Newdow's case was preceded by *Aronow v. U.S.*, 432 F.2d 242 (1970 U.S.).
5. See "Litigant Explains Why He Brought Pledge Suit." In 2000, Newdow sued Congress, then President George W. Bush, the state of California, and the Elk Grove School District, alleging that in-class recitation of the Pledge of Allegiance constituted a violation of First Amendment rights. The district court dismissed his complaint. *Newdow v. United States Congress,* 292 F.3d 597 (2000 E.D.Cal).
6. *Newdow v. United States Congress,* 292 F.3d 597 (2002 Ninth Circuit California).
7. Jessica Reaves, "Person of the Week: Michael Newdow," *Time* (28 June 2002), http://www.time.com/time/nation/article/0,8599,266658,00.html.
8. "Opinions on the Pledge of Allegiance Ruling," *CNN.com* (26 June 2002), http://articles.cnn.com/2002-06-26/justice/pledge.reax.quotes_1_pledge-newdow-judge-alfred-t-goodwin?_s=PM:LAW.
9. See the discussion in T. Jeremy Gunn, "Religious Freedom and *Laïcité:* A Comparison of the United States and France," *Brigham Young University Law Review* (Summer 2004), pp. 423–24, 496.
10. *Elk Grove Unified School District v. Newdow,* 542 U.S. 1, 124 S. Ct. 2301 (2004).
11. For examples of justices who proceeded to criticize the substantive as-

pects of Newdow's "under God" litigation, see William Rehnquist, *Elk Grove Unified School District v. Newdow,* 542 U.S. 1, 32 (2004), Sandra Day O'Connor (36, 44), and Clarence Thomas (53).

12. Caron, "*Laïcité* and Secular Attitudes in France," in Kosmin and Keysar, *Secularism and Secularity,* p. 113. Gunn, "Religious Freedom and *Laïcité,*" p. 428.
13. Thomas Jefferson, "84. A Bill for Punishing Disturbers of Religious Worship and Sabbath Breakers," in *The Papers of Thomas Jefferson. Vol. 2: 1777 to 18 June 1779, Including the Revisal of the Laws, 1776–1786,* edited by Julian Boyd (Princeton, NJ: Princeton University Press, 1950), p. 555. Also see Richard Bowser and Robin Muse, "Historical Perspectives on Church and State," in Duncan and Jones, *Church-State Issues in America Today,* vol. 1, p. 45.
14. Hutson, *Forgotten Features of the Founding,* pp. 63, 159.
15. Ibid., p. 63.
16. Ibid., p. 160.
17. Hamburger, *Separation of Church and State,* p. 162.
18. Hutson, *Forgotten Features of the Founding,* p. 173. Though Jefferson was opposed to the practice, see Hamburger, *Separation of Church and State;* Vincent Phillip Muñoz, "James Madison's Principle of Religious Liberty," *American Political Science Review* 97 (2003), p. 28. On some church-state trespasses of the younger Jefferson, see Dreisbach, *Thomas Jefferson and the Wall of Separation,* pp. 57–59.
19. James Madison, "By the President of the United States of America: A Proclamation," in *A Compilation of the Messages and Papers of the Presidents, 1789–1897,* vol. 1, edited by James Richardson (n.p., 1898), p. 513. For other proclamations by Madison, see pp. 532, 558, 560.
20. Muñoz, "James Madison's Principle of Religious Liberty," p. 28. Madison ruminated on these transgressions later in his life in "A Detached Memorandum," in Church, *The Separation of Church and State,* pp. 131–44.
21. Italics mine. James Madison, "Private. To. Rev.—— Adams," in *The Writings of James Madison. Vol. 9: 1819–1836* (New York: G. P. Putnam's Sons, 1910), p. 484.
22. Dreisbach, "'Sowing Useful Truths and Principles,'" p. 497.
23. *Zorach v. Clauson,* 343 U.S. 306, 312 (1952).
24. Italics mine. *Lemon v. Kurtzman,* 403 U.S. 602 (1971).
25. *Gillette v. United States,* 401 U.S. 437, 450 (1971).

26. "Loi de 9 décembre 1905 concernant la séparation des Églises et de l'État," *Légifrance.gouv.fr,* http://www.legifrance.gouv.fr/affichTexte.do?cidTexte=LEGITEXT000006070169&dateTexte=20101206.
27. "Constitution de la République française du 4 octobre 1958," Art. I, Assemblée Nationale, http://www.assemblee-nationale.fr/english/8ab.asp.
28. John Esposito, "Islam and Secularism in the Twenty-first Century," in Esposito and Tamimi, *Islam and Secularism in the Middle East,* p. 9.
29. Nikki Keddie, "Trajectories of Secularism in the West and the Middle East," *Global Dialogue* 6 (2004), p. 23.
30. Bernard Stasi, *Laïcité et République* (Paris: La Documentation Française, 2004), p. 129.
31. "Loi n 2004–228 du 15 mars 2004 encadrant, en application du principe de laïcité, le port de signes, ou de tenues manifestant une appartenance religieuse dans les écoles, collèges, et lycées publics," *Sénat,* http://www.senat.fr/dossier-legislatif/pjl03-209.html. The law has been implemented via the national education code (Art. L. 141-5-1). "Circulaire du 18 mai 2004 relative à la mise en oeuvre de la loi n 2004-228 du 15 mars 2004 encadrant, en application du principe de laïcité, le port de signes, ou de tenues manifestant une appartenance religieuse dans les écoles, collèges, et lycées publics," Art. II, *Légifrance.gouv.fr,* http://www.legifrance.gouv.fr/affichTexte.do?cidTexte=JORFTEXT000000252465&dateTexte=.
32. "Nul ne peut, dans l'espace public, porter une tenue destinée à dissimuler son visage"; "Loi n 2010-1192 du 11 octobre 2010 interdisant la dissimulation du visage dans l'espace public," Art. 1, *Légifrance.gouv.fr,* http://legifrance.gouv.fr/affichTexte.do?cidTexte=JORFTEXT000022911670&categorieLien=id.
33. Editorial, "The Taliban Would Applaud," *New York Times* (26 January 2010), http://www.nytimes.com/2010/01/27/opinion/27wed2.html>. Linda Chavez, "Banning the Burqa: Veil Is Anti-Woman," *New York Post* (22 May 2010), http://www.nypost.com/p/news/opinion/opedcolumnists/banning_the_burqa_QLNZArwCXHYohKXszTSSeJ. Eleanor Beardsley, "France's Burqa Ban Adds to Anti-Muslim Climate," *NPR* (11 April 2011), http://www.npr.org/2011/04/11/135305409/frances-burqa-ban-adds-to-anti-muslim-climate.
34. Editorial, "The Taliban Would Applaud." Chavez, "Banning the Burqa."

35. Notably among those in the religious liberty lobby. See, for example, the Beckett Fund's opposition to the idea: http://www.becketfund.org/index.php/case/96.html/print/.
36. Not that there were that many or perhaps *any* atheists in their time, but that is a story for another day. On this issue see Lucien Febvre, *The Problem of Unbelief in the Sixteenth Century: The Religion of Rabelais,* translated by Beatrice Gottlieb (Cambridge, MA: Harvard University Press, 1982); Alan Kors, *Atheism in France, 1650–1729. Vol. 1: The Orthodox Sources of Disbelief* (Princeton, NJ: Princeton University Press, 1990); Michael Buckley, *At the Origins of Modern Atheism* (New Haven, CT: Yale University Press, 1987). Suffice it to say that this body of work questions whether there were any atheists prior to the eighteenth century. In any case, atheists were very rare until the nineteenth century.
37. Baron Thiry d'Holbach, *Le bon sens, ou idées naturelles opposées aux idées surnaturelles* (Paris: Éditions Rationalistes, 1971), p. 11.
38. James Thrower, *Western Atheism: A Short History* (Amherst, NY: Prometheus Books, 2000), p. 106.
39. Rousseau, *The Social Contract,* p. 187.
40. Ibid., p. 53.
41. Voltaire, *Letters on England,* translated by Leonard Tanock (New York: Penguin, 1985), p. 41.
42. On this point see Ahmet Kuru, *Secularism and State Policies Toward Religion: The United States, France, and Turkey* (Cambridge: Cambridge University Press, 2009), pp. 23–25.
43. Henri Pena-Ruiz, *Qu'est-ce que la laïcité?* (Paris: Gallimard, 2003), p. 135. Also see Gunn, "Religious Freedom and *Laïcité,*" p. 433.
44. Cited in Eleanor Marx Aveling, *History of the Commune of 1871* (New York: Monthly Review Press, 1967), p. iii.
45. "Title I. Surveillance of the Exercise of Worship. Preliminary and General Provision," in Anderson, *The Constitutions and Other Select Documents,* p. 141.
46. "The Concordat, September 10, 1801–April 8, 1802," in Anderson, *The Constitutions and Other Select Documents,* p. 297.
47. Acts of brutalization were aimed not at members of the political or religious class. Rather, it was an immense population of enslaved persons of African descent who suffered its irrationalities.

48. Rousseau, *The Social Contract,* p. 88.
49. "Constitution of 1791," in Anderson, *The Constitutions and Other Select Documents,* p. 59–60. The 1789 text was incorporated in the constitution of 1791.
50. Gunn, "Religious Freedom and *Laïcité,*" pp. 466–68.
51. In the view of Jean Baubérot, Article 10 was "the affirmation – in, it is true, a very moderate form – of liberty of conscience." Baubérot, "Brève histoire de la laïcité en France," in *La laïcité à l'épreuve: Religions et libertés dans le monde,* edited by Jean Baubérot (Paris: Universalis, 2004), p. 145.
52. Pena-Ruiz, *Qu'est-ce que la laïcité?,* p. 9.
53. John Leland, "The Rights of Conscience," in Church, *The Separation of Church and State,* p. 100. In the same volume Jefferson makes his famous aside about "twenty gods or, no god" in "Notes on the State of Virginia," ibid., p. 52.
54. *Wallace v. Jaffree,* 472 U.S. 38, 98 (1985).
55. Gunn, "Religious Freedom and *Laïcité,*" p. 419.
56. For this point see Jean Baubérot's *Histoire de la laïcité française* (Paris: Presses Universitaires de France, 2000), p. 20.
57. See, for example, Article III, "Massachusetts Constitution of 2 March 1780," in Oscar Handlin and Mary Handlin (eds.), *The Popular Sources of Political Authority: Documents on the Massachusetts Constitution of 1780* (Cambridge, MA: Belknap Press of Harvard University Press, 1966), pp. 441–72; Preamble, Constitution of New York of April 20, 1777, in *Journals of the Provincial Congress, Provincial Convention Committee of Safety, and Council of Safety of the State of New York, 1775, 1776, 1777,* vol. 1 (Albany, NY: Thurlow Weed, Printer to the State, 1842), pp. 892–98; Sect. 36, Constitution of Pennsylvania – September 28, 1776, in Francis Newton Thorpe (ed.), *The Federal and State Constitutions, Colonial Charters, and Other Organic Laws of the States, Territories, and Colonies Now or Heretofore Forming the United States of America* (Washington, DC : U.S. Government Printing Office, 1909).
58. "House of Representatives, Amendments to the Constitution," document 11 in *Annals of Congress: The Debates and Proceedings in the Congress of the United States. Vol. 5: Bill of Rights* (Washington, DC:

Gales & Seaton, 1834–1856), http://press-pubs.uchicago.edu/founders/documents/bill_of_rightss11.html.

59. "Decree upon Religion, February 21, 1975," in Anderson, *The Constitutions and Other Select Documents,* p. 139.
60. "Concordat," in *OSV's Encyclopedia of Catholic History,* edited by Matthew Bunson (Huntington, IN: Our Sunday Visitor, 2004), p. 240.
61. Jean Baubérot, "Two Thresholds of *Laïcization,*" in Bhargava, *Secularism and Its Critics,* p. 102.
62. Ibid., p. 102.
63. Baubérot, "Two Thresholds of *Laïcization,*" pp. 102–3.
64. Baubérot, *Histoire de la laïcité française,* p. 24.
65. "General Provision for all the Protestant Communions," in Anderson, *The Constitutions and Other Select Documents,* p. 307.
66. Ibid.
67. Olivier Roy, *Secularism Confronts Islam,* translated by George Holoch (New York: Columbia University Press, 2007), p. 33.
68. Ibid., p. 41. Also see p. 3.
69. John Bowen, *Why the French Don't Like Headscarves* (Princeton, NJ: Princeton University Press, 2007), pp. 15, 18, 22.
70. Translation mine. The original reads, "La République assure la liberté de conscience. Elle garantit le libre exercice des cultes sous les seules restrictions édictées ci-après dans l'intérêt de l'ordre public." "Loi du 9 décembre 1905 concernant la séparation des Églises et de l'État," Art. 1, *Légifrance.gouv.fr,* http://www.legifrance.gouv.fr/affichTexte.do?cidTexte=LEGITEXT000006070169&dateTexte=20101206.
71. "La République ne reconnaît, ne salarie, ni ne subventionne aucun culte." Ibid., Art. II.
72. "Pourront toutefois être inscrites auxdits budgets les dépenses relatives à des services d'aumônerie et destinées à assurer le libre exercice des cultes dans les établissements publics tels que lycées, collèges, écoles, hospices, asiles, et prisons." Ibid.
73. See "Separation Law" (extracts) in Maclear, *Church and State in the Modern Age: A Documentary History,* pp. 304–10.
74. Bruce Crumley, "Scientology Trial in France: Can a Religion Be Banned?" *Time* (28 May 2009), http://www.time.com/time/world/article/0,8599,1901373,00.html. Christopher Beam, "Cult Busters:

How Governments Decide Whether a Religion Is Real or Not," *Slate* (28 October 2009), http://www.slate.com/id/2233850/. The report of the Parliamentary Commission on Cults in France (1995) gives a list of 172 "cults" that have been deemed dangerous. Among them are Jehovah's Witnesses and Hare Krishnas. M. Jacques Guyar, "Rapport fait au nom de la commission d'enquête (1) sur les sects," Assemblée Nationale (22 December 1995), http://www.assemblee-nationale.fr/rap-enq/r2468.asp.

75. Bowen, *Why the French Don't Like Headscarves*, p. 16.
76. Valentine Zuber, "La commission Stasi et les paradoxes de la laïcité française," in Baubérot, *La laïcité à l'épreuve*, p. 31. The practice is discussed by Sarkozy in *La république, les religions, l'ésperance*, pp. 23–24.
77. Kuru, *Secularism and State Policies*, p. 109.
78. Susan Sachs, "France's Army Embraces Its Muslim Soldiers," *Globe and Mail* (14 September 2009), http://www.theglobeandmail.com/news/world/frances-army-embraces-its-muslim-soldiers/article1287552/.
79. As the current French president, Nicolas Sarkozy, who was himself once minister of religions, has remarked, "*Laïcité* is not the enemy of religion. Quite the contrary. *Laïcité* is the guarantee that each person is able to believe and live his faith." Sarkozy, *La république, les religions, l'ésperance*, p. 15.
80. "Atheist Group Sues over Cross at Sept. 11 Museum," *Wall Street Journal* (27 July 2011), http://online.wsj.com/article/APff6ef67bd49547e4a93073b5422ccea8.html.

4. Does Secularism Equal Atheism?

Epigraph: George Jacob Holyoake, *English Secularism: A Confession of Belief* (Chicago: The Open Court Publishing Company, 1896).

1. Compare, for example, entries of the anti-metaphysical variety, such as P. DeLetter, "Secularism," in *Encyclopedic Dictionary of Religion. Vol. O–Z*, edited by Paul Kevin Meagher et al. (Washington, DC: Corpus Publications, 1979), p. 3241, with a definition more in line with Holyoake's ethical/existential approach, and Eric Waterhouse, *Encyclopedia of Religion and Ethics*, edited by James Hastings (New York: Charles Scribner's Sons, 1921), pp. 347–50. A more inclusive (and in

my opinion more accurate) definition is offered by David Tribe, "Secularism," in *The New Encyclopedia of Unbelief*, edited by Tom Flynn (Amherst, NY: Prometheus Books, 2007), pp. 700–703.

2. Lisa Miller, "In Defense of Secularism," *Newsweek*, vol. 151, no. 8 (25 February 2008), http://www.newsweek.com/2008/02/16/in-defense-of-secularism.html.
3. Secular Coalition for America, "About the Secular Coalition for America," http://www.secular.org/node/53.
4. Ibid.
5. Secular Student Alliance, "About the Secular Student Alliance," http://www.secularstudents.org/about.
6. Membership numbers for these groups are difficult to find, but the size of one of them, American Atheists, stood at about twenty-two hundred members a few years ago. Eugenie Carol Scott, *Evolution vs. Creationism* (Berkeley: University of California Press, 2005), p. 66.
7. Penny Edgell and Joseph Gerteis, "Atheists as 'Other': Moral Boundaries and Cultural Membership in American Society," *American Sociological Review* 71 (2006), pp. 211–34.
8. Pew Forum on Religion and Public Life, "Public Expresses Mixed Views of Islam, Mormonism," Pew Research Center (25 September 2007), http://pewresearch.org/pubs/602/public-expresses-mixed-views-of-islam-mormonism.
9. Victor Stenger, *The New Atheism: Taking a Stand for Science and Reason* (Amherst, NY: Prometheus Books, 2009), p. 25.
10. Gary Wolf, "The Church of the Non-Believers," *Wired*, vol. 14, no. 1 (November 2006), http://www.wired.com/wired/archive/14.11/atheism.html. Simon Hooper, "The Rise of the 'New Atheists,'" *CNN.com* (8 November 2006), http://articles.cnn.com/2006-11-08/world/atheism.feature_1_new-atheists-new-atheism-religion?_s=PM:WORLD.
11. Sam Harris, *The End of Faith: Religion, Terror, and the Future of Reason* (New York: W. W. Norton, 2005), p. 45. It is this belief that led Harris to aver – in the face of everything we know about the history of religion – that "religious moderation . . . does not permit anything very critical to be said about religious literalism" (p. 20). Harris may want to read up on the history of the documentary hypothesis in biblical studies, the liberal Protestant clergy who espoused it, and the heresy trials that ensued. See Jacques Berlinerblau, "'Poor Bird, Not

Knowing Which Way To Fly': Biblical Scholarship's Marginality, Secular Humanism, and the Laudable Occident," *Biblical Interpretation* 4 (2002), pp. 267–304. On the New Atheists' unwillingness "to take a benign view of moderate religion," see Stenger, *The New Atheism*, p. 13.

12. Richard Dawkins, *The God Delusion* (Boston: Houghton Mifflin, 2006), pp. 301–8.
13. George Jacob Holyoake, *The Origin and Nature of Secularism* (London: Watts and Co., 1896), pp. 51–52. He makes the same point in a work that is almost but not completely identical to this one, titled, *English Secularism: A Confession of Belief*, pp. 46–47. Elsewhere, Holyoake sets the date at 1852 or 1853; George Jacob Holyoake and Charles Bradlaugh, *Secularism, Scepticism, and Atheism: Verbatim Report of the Proceedings of a Two Nights' Public Debate* (London: Austin and Co., 1870), p. 1.
14. George Jacob Holyoake, *The Principles of Secularism* (London: Austin & Co., 1871). Throughout the text secularism is variously associated with utilitarianism (p. 11), freethought (pp. 9–10, 38), service for others (p. 11), positivism (p. 11), naturalism (p. 33), an emphasis on science (p. 12), this-worldliness (27), sincerity (12), materialism (11), something other than theism or atheism (p. 11), a form of religiousness (p. 12), the free search for truth (p. 14), and free speech (pp. 14–15), among other concepts.
15. Holyoake provides a doleful list of all of those freethinkers who were tried, imprisoned, or fined (p. 15) in *The Origin and Nature of Secularism.* He himself spent half a year in Gloucester Gaol in 1842. Joseph McCabe, *Life and Letters of George Jacob Holyoake*, vol. 1, edited by Charles William and Frederick Goss (London: Watts & Co., 1908), p. 83.
16. George Jacob Holyoake, *A Short and Easy Method with the Saints* (London: Hetherington, 1843), p. 6. The date of 1843 is not entirely certain.
17. Ibid., p. 9. The circumstances of his arrest are as follows: Having "carefully avoided" mentioning religion during his lecture, Holyoake responded to a query from an irate clergyman by saying, "I am not religious – my creed is to have no creed. But what do I hear? That morality cannot exist without religion? Preposterous! Religion in my opinion has ever poisoned the fountain-springs of morality. Connect them to-

gether! Hark ye! Morality alone is lovely." McCabe, *Life and Letters of George Jacob Holyoake,* vol. 1, p. 64.

18. Holyoake, *A Short and Easy Method,* p. 27.
19. Ibid., p. 28. See David Berman, *A History of Atheism in Britain: From Hobbes to Russell* (New York: Croom Helm, 1988), p. 213; p. 209 for Holyoake's professions of atheism.
20. Edward Royle, *Victorian Infidels: The Origins of the British Secularist Movement, 1791–1866* (Manchester, UK: Manchester University Press, 1974), pp. 72, 158, 280. Others have noted that the famed infidel sometimes veered into pantheism. See Lee Grugel, *George Jacob Holyoake: A Study in the Evolution of a Victorian Radical* (Philadelphia: Porcupine Press, 1976), p. 95.
21. Royle, *Victorian Infidels,* pp. 72, 158, 280.
22. Holyoake and Bradlaugh, *Secularism, Scepticism, and Atheism,* p. 47.
23. Of interest is how so many major Victorian infidels ended their lives in the domain of theism. Robert Owen was channeling spirits at the end of his life. Richard Carlile became a deist. Royle, *Victorian Infidels,* p. 42. Charles Southwell recanted, published an essay called *The Impossibility of Atheism* in 1852, and came to Christ. See Berman, *A History of Atheism in Britain,* pp. 212, 213, 206–7. Even Charles Bradlaugh, whom we will soon meet, seemed less than a "pure" atheist. David Berman notes that there is "a certain tension or vacillation in Bradlaugh's atheism." Ibid., p. 215. He often danced around confessions of pure nonbelief. Looking back on his life, an older Holyoake did not describe Bradlaugh as an "absolute atheist." Berman finds this omission meaningful and, given Bradlaugh's reputation as the arch-atheist of the era, it is all quite fascinating. Ibid., p. 215. On Bradlaugh, also see Walter Arnstein, *The Bradlaugh Case: Atheism, Sex, and Politics Among the Late Victorians* (Columbia: University of Missouri Press, 1983), p. 11.
24. On this point, see Bruce Hunsberger and Bob Altemeyer, *Atheists: A Groundbreaking Study of America's Nonbelievers* (Amherst, NY: Prometheus Books, 2006), p. 53.
25. Holyoake, *The Origin and Nature of Secularism,* p. 51.
26. The book published in 1871 as *The Principles of Secularism* (London: Austin & Co., 1871) refers to a version published sixteen years earlier (p. 5).

27. Holyoake, *The Principles of Secularism*, p. 11.
28. Holyoake, *English Secularism*, p. 70. Also see Holyoake and Bradlaugh, *Secularism, Scepticism, and Atheism*, p. 31.
29. Ibid., p. 35.
30. Ibid.
31. Holyoake, *The Principles of Secularism*, p. 27.
32. McCabe, *Life and Letters of George Jacob Holyoake*, vol. 1, p. 208.
33. To this we will add that Holyoake placed an emphasis on the importance of free inquiry (*Origin and Nature of Secularism*, pp. 10, 14), even poking fun at the Protestant hypocrisy on this point. *Principles of Secularism*, p. 40.
34. David Tribe, *President Charles Bradlaugh, M.P.* (London: Elek Books, 1971), p. 10. Also see Janet Courtney, *Freethinkers of the Nineteenth Century* (London: Chapman and Hall, 1920), p. 97.
35. And his advocacy of "neo-Malthusianism," or what we would now call birth control, did not make him any less outrageous to his fellow Victorians. On these points see Walter Arnstein, *The Bradlaugh Case*, p. 55.
36. Holyoake and Bradlaugh, *Secularism, Scepticism, and Atheism*, p. viii.
37. David Stewart Nash, "Bradlaugh, Charles (1833–1891)," in *The Dictionary of Nineteenth-Century British Philosophers. Vol. 1: A–H*, edited by Gavin Budge et al. (Bristol, UK: Thoemmes Press, 2002), pp. 129–33. Also see Jim Herrick, "National Secular Society," in Flynn, *The New Encyclopedia of Unbelief*, pp. 556–57. Then again, biographers show him to be more thoughtful in private life. He was deeply knowledgeable about theology and religious affairs and even maintained warm relations with liberal Christians. David Tribe, "Charles Bradlaugh," in *Icons of Unbelief: Atheists, Agnostics and Secularists*, edited by S. T. Joshi (Westport, CT: Greenwood Press, 2008), p. 9.
38. For a breathy, ringside-style account, see Tribe, *President Charles Bradlaugh*, p. 210. On his six-year wait, see Tribe, "Bradlaugh, Charles," in Flynn, *The New Encyclopedia of Unbelief*, pp. 150–52.
39. See, for example, Jim Herrick, "Holyoake, George Jacob," in Flynn, *The New Encyclopedia of Unbelief*, pp. 396.
40. "Dawkins Attack on Peter Kay Is Not Very Christian," *Theatre* blog, *The Guardian* (8 March 2007), http://www.guardian.co.uk/stage/theatreblog/2007/mar/08/dawkinsattackonpeterkayis. Dekka Aitkenhead,

"People Say I'm Strident," *The Guardian* (25 October 2008), http://www.guardian.co.uk/science/2008/oct/25/richard-dawkins-religion-science-books.

41. Holyoake and Bradlaugh, *Secularism, Scepticism, and Atheism*, p. 70.
42. Ibid., p. 38.
43. Ibid., p. 41.
44. Ibid., p. iii.
45. Quoted in *Principles of Secularism*, p. 8.
46. David Tribe notes that *secularism* "is the term by which unbelief was generally known in the English-speaking world for most of the second half of the nineteenth century." "Secularism," in Flynn, *The New Encyclopedia of Unbelief*, p. 701. This seems to apply to England and not the United States. In the preface to his work *English Secularism* of 1896 he reiterated, "The new form of free thought known as English Secularism does not include either Theism or Atheism." Holyoake, *English Secularism*, unpaginated author's preface.
47. Holyoake, *The Origin and Nature of Secularism*, p. 42.
48. Holyoake and Bradlaugh, *Secularism, Scepticism, and Atheism*, pp. v, 69.
49. Ibid., p. 16.
50. Ibid., p. 64.
51. Ibid., p. vii.
52. Mitchell Landsberg, "Religious Skeptics Disagree on How Aggressively to Challenge the Devout," *Los Angeles Times* (10 October 2010), http://articles.latimes.com/010/oct/10/local/la-me-humanists-20101010.
53. Secularism neither asserts nor denies theism, or atheism, for that matter. Holyoake, *The Origin and Nature of Secularism*, p. 42.
54. Ibid., p. 46; Bradlaugh responds to this (p. 57).
55. Holyoake and Bradlaugh, *Secularism, Scepticism, and Atheism*, pp. 13, 24.
56. Bradlaugh also infers as much. Ibid., p. 4.
57. Ibid., p. 73.
58. "The National Secular Society is a minority organization unable to attract much public attention or a significant number of members." Graeme Smith, *A Short History of Western Secularism* (London: I. B. Tauris, 2008), pp. 173, 172–75. On the declining membership of the NSS and its overall history, see Edward Royle, *Radicals, Secularists,*

and Republicans: Popular Freethought in Britain, 1866–1915 (Manchester, UK: Rowman and Littlefield, 1980), pp. 132–36.

59. Basharat Peer, "Zero Tolerance and Cordoba House," *FT.com* (13 August 2010), http://www.ft.com/cms/s/2/bf1110d8-a5b0-11df-a5b7-00144feabdc0.html.
60. Ibid.
61. Michael Barbaro and Marjorie Connelly, "New Yorkers Divided over Islamic Center, Poll Finds," *New York Times* (2 September 2010), http://www.nytimes.com/2010/09/03/nyregion/03poll.html?ref=opinion.
62. Center for Inquiry, "CFI Board Accepts Paul Kurtz's Resignation" (18 May 2010), http://www.centerforinquiry.net/news/center_for_inquiry_board_statement.
63. Paul Kurtz, "Apologia," http://paulkurtz.net/apologia.html; Landsberg, "Religious Skeptics Disagree on How Aggressively to Challenge the Devout"; Mark Oppenheimer, "Closer Look at Rift Between Humanists Reveals Deeper Division," *New York Times* (1 October 2010), p. A12.
64. Center for Inquiry, "CFI Releases Statement on Ground Zero Controversy" (29 August 2010), http://www.centerforinquiry.net/news/statement_on_ground_zero_controversy/.
65. Ibid.
66. "Prominent American Muslims Denounce Terror Committed in the Name of Islam: Transcript of CBS's *60 Minutes* Interview on 30 September 2001," *Islam for Today,* http://www.islamfortoday.com/60minutes.htm.
67. For the text of the full interview, see "What Does It Take to Change the Relationship Between the West and the Muslim World?: A Public Lecture with Imam Feisal Abdul Rauf," Bob Hawke Prime Ministerial Centre (12 July 2005), http://www.unisa.edu.au/hawkecentre/events/2005events/Imam_transcript.asp.
68. Nat Hentoff, "Am I Also a Bigot?: Pols Clueless on the Ground Zero Mosque," *Jewish World Review* (25 August 2010), http://www.jewishworldreview.com/cols/hentoff082510.php3.
69. Website of Cordoba Initiative, "Frequently Asked Questions," http://www.cordobainitiative.org/?q=content/frequently-asked-questions.
70. "Hamas Leader Zahar: Muslims Must Build Mosque Near Ground Zero," *Ha'aretz* (16 August 2010), http://www.haaretz.com/news

/international/hamas-leader-zahar-muslims-must-build-mosque-near-ground-zero-1.308387.

71. Kenneth Vogel and Giovanni Russonello, "Latest Mosque Issue: The Money Trail," *Politico.com* (4 September 2010), http://www.politico.com/news/stories/0910/41767.html. Russell Goldman, "Islamic Center Backers Won't Rule Out Taking Funds from Saudi Arabia, Iran," *ABCNews.com* (18 August 2010), http://abcnews.go.com/US/Politics/islamic-center-backers-rule-taking-funds-saudi-arabia/story?id=11429998.
72. Paul Vitello, "Amid Rift, Imam's Role in Islam Center Is Sharply Cut," *New York Times* (14 January 2011), http://www.nytimes.com/2011/01/15/nyregion/15mosque.html.

5. *How Not to Be Secular*

Epigraph: Philip Roth, *I Married a Communist* (Boston: Houghton Mifflin, 1998).

1. Eric Marx, "Senator Urges Orthodox Leaders to Wage War on Secularism," *Jewish Daily Forward* (14 November 2003), http://www.forward.com/articles7061.
2. Jacqueline Maley, "Thank God We're Not All Atheists, Bishop Says," *Sydney Morning Herald* (2 April 2010), http://www.smh.com.au/national/thank-god-were-not-all-atheists-bishop-says-20100401-ri4q.html.
3. Nandy, "Closing the Debate on Secularism," in Needham and Rajan, *The Crisis of Secularism in India*, p. 111.
4. "Separation Law" (extracts), in Maclear, *Church and State in the Modern Age*, pp. 304–10.
5. Stalin's observation is quoted in an interview excerpt found in Wladyslaw Kania, *Bolshevism and Religion*, translated by R. M. Dowdall (New York: Polish Library, 1946), p. 16. The comment was made in 1928 and first published in Russian in 1938.
6. Karl Marx, "On the Jewish Question," in *The Marx-Engels Reader*, edited by Robert Tucker, 2nd ed. (New York: W. W. Norton, 1978), p. 28.
7. Quoted in Albert Boiter, *Religion in the Soviet Union: The Washington Papers*, vol. 78 (Beverly Hills, CA: Sage, 1980), p. 19. James Thrower

repeatedly makes the point that "Lenin added little, if anything, to the classical Marxist analysis of religion." *Marxist-Leninist 'Scientific Atheism' and the Study of Religion and Atheism in the USSR* (Berlin: Mouton, 1983), p. 116.

8. David Shub, *Lenin: A Biography* (Baltimore: Penguin, 1967), p. 36.
9. V. I. Lenin, "The Attitude of the Workers' Party to Religion," in *V. I. Lenin: Collected Works. Vol. 15: March 1908–August 1909* (Moscow: Progress Publishers, 1982), p. 402.
10. Mikhail Bakunin, *God and the State* (New York: Dover, 1970), p. 15. Friedrich Engels, "Emigrant Literature," in *Marx/Engels on Religion* (Moscow: Progress Publishers, 1985), p. 124.
11. V. I. Lenin, "Socialism and Religion," in *Religion* (New York: International Publishers, 1933), pp. 9–10. In this same 1905 essay Lenin opined that his program does not need to declare itself atheist and even argues that the religious proletariat should not be discouraged from joining the movement.
12. Marx and Engels, "The German Ideology," in Tucker, *The Marx-Engels Reader*, p. 154.
13. Sergei Hackel, "Union of Soviet Socialist Republics, Christianity," in *Religion in Politics: A World Guide*, edited by Stuart Mews (Chicago: St. James Press, 1989), p. 272.
14. Andrew Sorokowski, "Church and State, 1917–1964," in *Candle in the Wind: Religion in the Soviet Union*, edited by Eugene Shirley Jr. and Michael Rowe (Washington, DC: Ethics and Public Policy Center, 1989), p. 22.
15. Paul Froese, *The Plot to Kill God: Findings from the Soviet Experiment in Secularization* (Berkeley: University of California Press, 2008), p. 49.
16. Marc Szeftel, "Church and State in Imperial Russia," in *Russian Orthodoxy Under the Old Regime*, edited by Robert Nichols and Theofanis Stavrou (Minneapolis: University of Minnesota Press, 1978), pp. 136–37.
17. Ibid.
18. Sorokowski, "Church and State," in Shirley and Rowe, *Candle in the Wind*, p. 23.
19. Alexander Yakovlev, *A Century of Violence in Religious Russia*, trans-

lated by Anthony Austin (New Haven, CT: Yale University Press, 2002), p. 156.

20. And often the Party promoted smaller denominations as a means of undermining the Orthodox Church. On this point, see David Powell, *Antireligious Propaganda in the Soviet Union: A Study of Mass Persuasion* (Cambridge, MA: The MIT Press, 1975), p. 27. On the irony that the Soviets inadvertently "planned the seeds of religious pluralism" by focusing on the Orthodox faith, see Paul Froese, "Forced Secularization in Soviet Russia: Why an Atheistic Monopoly Failed," *Journal of Sociology and Social Anthropology* 43 (2004), p. 40.
21. Froese, "Forced Secularization," p. 35.
22. Philip Walters, "A Survey of Soviet Religious Policy," in *Religious Policy in the Soviet Union*, edited by Sabrina Ramet (Cambridge: Cambridge University Press, 1993), p. 3.
23. Powell, *Antireligious Propaganda*, p. 36.
24. Adam Jolles, "Stalin's Talking Museums," *Oxford Art Journal* 28 (2005), p. 432.
25. Ibid., pp. 446–47.
26. John Anderson, *Religion, State, and Politics in the Soviet Union and Successor States* (Cambridge: Cambridge University Press, 1994), p. 11.
27. Froese, "Forced Secularization," pp. 37–38. On the formation of the league, see Daniel Peris, *Storming the Heavens: The Soviet League of the Militant Godless* (Ithaca, NY: Cornell University Press, 1998), p. 44.
28. William Husband, "Soviet Atheism and Russian Orthodox Strategies of Resistance, 1917–1932," *The Journal of Modern History* 70 (1998), p. 74.
29. Walter Kolarz, *Religion in the Soviet Union* (New York: St. Martin's Press, 1961), p. 11.
30. Froese, "Forced Secularization," pp. 43, 46; Husband, "Soviet Atheism," p. 81; Peris, *Storming the Heavens*, pp. 27, 86. Of course, the public display of an embalmed and publicly displayed Lenin corpse was meant to insinuate his eternal existence.
31. Husband, "Soviet Atheism," p. 99; William Husband, *"Godless Communists": Atheism and Society in Soviet Russia, 1917–1932* (DeKalb: Northern Illinois University Press, 1999), p. 50.

32. Gerhard Simon, *Church, State, and Opposition in the U.S.S.R.* (Berkeley: University of California Press, 1974), p. 94.
33. Peris, *Storming the Heavens,* p. 94.
34. On these points generally see Jacques Berlinerblau, *The Secular Bible: Why Nonbelievers Must Take Religion Seriously* (Cambridge: Cambridge University Press, 2005).
35. Froese, "Forced Secularization," p. 48.
36. Ibid.
37. Simon, *Church, State, and Opposition,* p. 93.
38. Kolarz, *Religion in the Soviet Union,* p. 6.
39. Peris, *Storming the Heavens,* p. 88.
40. See Joan Delaney, "The Origins of Soviet Antireligious Organizations," in Marshall, *Aspects of Religion in the Soviet Union,* pp. 103–29; also Paul Gabel, *And God Created Lenin: Marxism vs. Religion in Russia, 1917–1929* (Amherst, NY: Prometheus Books, 2005), pp. 318–23.
41. On this point see Husband, "*Godless Communists,*" p. 37.
42. These examples taken from Donald Lowrie and William Fletcher, "Khrushchev's Religious Policy, 1959–1964," in Marshall, *Aspects of Religion in the Soviet Union,* pp. 131–55.
43. Froese, "Forced Secularization," p. 38.
44. Ibid.
45. Peter Prifti, "Albania—Towards an Atheist Society," in *Religion and Atheism in the U.S.S.R. and Eastern Europe,* edited by Bohdan Bociurkiw et al. (Toronto: University of Toronto Press, 1975), pp. 388–404.
46. "Decree on the Separation of Church and State of January 23, 1918," in Marshall, *Aspects of Religion in the Soviet Union,* pp. 437–38.
47. Amila Butorovic, "When Secularism Opposes Nationalism: The Case of the Former Yugoslavia," in *The Future of Secularism,* edited by T. N. Srinivasan (New Delhi: Oxford University Press, 2007), pp. 284–304.
48. On this point see William van den Bercken, *Ideology and Atheism in the Soviet Union* (Berlin: Mouton de Gruyter, 1989), pp. 65, 69, 91.
49. Gabel, *And God Created Lenin,* p. 135.
50. "Article 124, Constitution of the USSR of December 5, 1936," in Marshall, *Aspects of Religion in the Soviet Union,* p. 437. Joshua Rothenberg notes that Stalin's Constitution actually subtly annulled an earlier provision that permitted the right to propagate religion. Now Soviets

had a right "to profess religion, but not to propagate it." "The Legal Status of Religion in the Soviet Union," in Marshall, *Aspects of Religion in the Soviet Union*, pp. 65, 82.

51. "Article 124," in Marshall, *Aspects of Religion in the Soviet Union*, p. 437.
52. Simon, *Church, State, and Opposition*, p. 64.
53. Albert Boiter, *Religion in the Soviet Union*, p. 16. Also see Froese, "Forced Secularization," p. 40; Peris, *Storming the Heavens*, 6; Sorokowski, "Church and State," in Shirley and Rowe, *Candle in the Wind*, pp. 28–29.
54. Milton Himmelfarb, "Church and State: How High a Wall?" *Commentary* (July 1966), p. 25.
55. Merriman, "Finland," in *Religion and the State*, pp. 171–72.
56. Ibid., "Norway," pp. 258–59; "Denmark," pp. 155–56. See Phil Zuckerman, *Society Without God: What the Least Religious Nations Can Tell Us About Contentment* (New York: New York University, 2008).
57. Holyoake, *English Secularism*, p. 116.
58. Bryan Le Beau, *The Atheist: Madalyn Murray O'Hair* (New York: New York University Press, 2003), p. 273.

6. The "Rise" of American Secularism and the Secularish

Epigraph: Ulysses S. Grant, "Address to the Army of Tennessee," September 1875.

1. The Secular Coalition of America (which, we noted earlier, appears to be an atheist group speaking in secularism's name) once boasted that there were 63 million non-theists in the United States. The present Secular Coalition for America webpage devoted to constituency claims that there were 3.6 million atheists or agnostics in the United States in 2008. Secular Coalition for America, "Who Does the Secular Coalition for America Represent?," http://www.secular.org/constituency.html. The previous iteration of this page was based on a Harris Interactive poll in which 9 percent of respondents claimed to "believe there is no God" and 12 percent said they were "not sure whether or not there is a God." The SCA thereby came up with a figure of 63 million non-theists. That claim is no longer on their website. For a further discus-

sion of this, see Berlinerblau, *Thumpin' It,* p. 173, n73. This claim has been batted around the atheist blogosphere, as evidenced by the webpage of the International Humanist and Ethical Organization, "Secularism in the USA," http://www.iheu.org/secularism-usa.

2. Dennett, "The Bright Stuff."
3. Ibid.
4. For a complete discussion of this phenomenon, see Berlinerblau, *Thumpin' It.*
5. Susan Jacoby, *Freethinkers: A History of American Secularism* (New York: Metropolitan Books, 2004), pp. 6–7.
6. Barry A. Kosmin, Egon Mayer, and Ariela Keysar, *American Religious Identification Survey (ARIS) 2001,* The Graduate Center of the City University of New York, p. 13, http://www.gc.cuny.edu/faculty/research_studies/aris.pdf.
7. Ibid. Fascinatingly, these "no religion" groups included the following categories: "atheist," "agnostic," "humanist," "secular," and "no religion." Of these, the "no religion" category was overwhelmingly the largest (93.2 percent) of the "unchurched."
8. The problem may have been caused by the authors of the survey report rather than by Jacoby and others, for it was the demographers Kosmin, Mayer, and Keysar who in their introduction equated "secular" with "unchurched": "Often lost amidst the mesmerizing tapestry of faith groups that comprise the American population is also a vast and growing population of those without faith. They adhere to no creed nor choose to affiliate with any religious community. These are the seculars, the unchurched, the people who profess no faith in any religion," *ARIS 2001,* p. 5. For reasons already noted, speaking about these different categories as if they amounted to the same thing is inaccurate.

   A nearly identical error was made by Daniel Dennett. He was working from Pew Forum data indicating that roughly 10 percent of respondents claimed to be atheists (1 percent) or agnostics (2 percent), or to have "no preference" for a religion (8 percent). Adding up these percentages, Dennett concluded that a tenth of the American population, or 27 million people, were his fellow travelers, eager secularists with bayonets at the ready.
9. A bevy of articles of late has pointed to a rise in the number of nonaffiliated and nonbelieving Americans, though they make the same error

in generalizing across groups. Laurie Goodstein, "More Atheists Shout It from the Rooftops," *New York Times* (26 April 2009), p. A1, http://www.nytimes.com/2009/04/27/us/27atheist.html.

10. The Pew Forum on Religion and Public Life, "Faith on the Hill: The Religious Composition of the 112th Congress," The Pew Research Center (5 January 2011), http://pewforum.org/Government/Faith-on-the-Hill–The-Religious-Composition-of-the-112th-Congress.aspx. See also Jacques Berlinerblau, "No Atheists in Congress: Why?" *Brainstorm* blog, *Chronicle of Higher Education* (18 January 2011), http://chronicle.com/blogs/brainstorm/no-atheists-in-congress-why/31230. That Peter Stark of California did not respond as an atheist is something of a curiosity.
11. An interesting analysis can be found in Richard Cimino and Christopher Smith, "Secular Humanism and Atheism Beyond Progressive Secularism," *Sociology of Religion* 68 (2007), pp. 407–24.
12. President Obama said, for instance, "We are a nation of Christians and Muslims, Jews and Hindus, and nonbelievers. We are shaped by every language and culture, drawn from every end of this Earth." "Barack Obama's Inaugural Address," transcript, *New York Times* (20 January 2009), http://www.nytimes.com/2009/01/20/us/politics/20text-obama.html?pagewanted=all.
13. Congressman Peter Stark had, however, previously denied the existence of a supreme being. Why he didn't register as an atheist in the Pew survey is anybody's guess. Carla Marinucci, "Stark's Atheist Views Break Political Taboo," *SFGate.com* (14 March 2007), http://articles.sfgate.com/2007-03-14/news/17235967_1_atheist-secular-coalition-political-suicide.
14. Paul Kane, "Rival Is Church Official: Sen. Dole Makes Issue of 'Godless' Group," *Washington Post* (31 October 2008), http://www.washingtonpost.com/wp-dyn/content/article/2008/10/30/AR2008103004432.html.
15. "Dole Challenger Irate over Suggestion She Is 'Godless,'" *CNNPolitics.com* (30 October 2008), http://edition.cnn.com/2008/POLITICS/10/30/dole.ad/.
16. "Senate Candidate Files Lawsuit over 'Godless' Ad," *The Caucus* blog, *New York Times* (30 October 2008), http://thecaucus.blogs.nytimes.com/2008/10/30/senate-candidate-files-lawsuit-over-godless-ad/.

17. Ibid. The lawsuit was eventually dropped.
18. On the lack of popularity of atheists, see Phil Zuckerman, "Atheism, Secularity, and Well-being: How the Findings of Social Science Counter Negative Stereotypes and Assumptions," *Sociology Compass* 3 (2009), p. 949.
19. Freedom from Religion Foundation, "Welcome Members to the Freedom from Religion Foundation Forum!," http://ffrf-forum.org/. Parenthetically, lack of numbers has been endemic in the history of American freethought. See Warren's discussion of the American Secular Union's lack of membership, where he notes it declined from a height of fifty thousand members in 1900 to "a state of impotence from which it never emerged" ten years later. Sidney Warren, *American Freethought, 1860–1914* (New York: Columbia University Press, 1943), p. 175.
20. Tony Hileman, "Remarks by Tony Hileman" at a National Day of Reason event, Washington, DC (2003), http://www.nationaldayofreason.org/hileman03.html.
21. Scott, *Evolution vs. Creationism,* p. 66. For more on the relatively small numbers of secularists, see Le Beau, *The Atheist,* p. 303.
22. Le Beau, *The Atheist,* p. 17.
23. Ibid., p. 2.
24. Ibid., pp. 133–36. See also Ann Rowe Seaman, *America's Most Hated Woman: The Life and Gruesome Death of Madalyn Murray O'Hair* (New York: Continuum, 2005), p. 350. For the *Life* magazine article, see Jane Howard, "The Most Hated Woman in America," *Life,* vol. 56, no. 25 (19 June 1964), pp. 91–94.
25. *Abington Sch. Dist. v. Schempp,* 374 U.S. 203 (1962).
26. Le Beau, *The Atheist,* p. 135.
27. Ibid., p. 151.
28. Ibid., p. 249.
29. See Seaman, *America's Most Hated Woman.*
30. Ibid.
31. On Ingersoll, see, for example, Tom Flynn and Roger Greeley, "Ingersoll, Robert Green," in Flynn, *The New Encyclopedia of Unbelief,* p. 423. Orvin Larson, *American Infidel: Robert G. Ingersoll* (New York: Citadel Press, 1962).
32. For some data on the size of the freethought movement in the 1830s,

see Albert Post, *Popular Freethought in America, 1825–1850* (New York: Columbia University Press, 1943), pp. 192–93.

33. Flynn and Greeley, "Ingersoll, Robert Green," in Flynn, *The New Encyclopedia of Unbelief,* p. 424.
34. Le Beau, *The Atheist,* p. 6.
35. Evelyn Kirkley, *Rational Mothers and Infidel Gentlemen: Gender and American Atheism, 1865–1915* (Syracuse, NY: Syracuse University Press, 2000), p. xiv.
36. Ingersoll was elected president of the National Liberal League in 1877. Larson, *American Infidel,* p. 144.
37. Flynn and Greeley, "Ingersoll, Robert Green," in Flynn, *The New Encyclopedia of Unbelief,* p. 426; Cooke, "National Liberal League," ibid., p. 555.
38. "The resolution was finally adopted, and Colonel Ingersoll resigned his office of vice-president in the League, and never acted with it again until the League dropped all side issues, and came back to first principles – the enforcement of the Nine Demands of Liberalism." Robert G. Ingersoll, "The Circulation of Obscene Literature," in *The Works of Robert G. Ingersoll,* vol. 12, edited by C. P. Farrell (New York: Dresden Publishing Co., 1909), p. 230.
39. Samuel Porter Putnam, *Four Hundred Years of Freethought* (New York: The Truth Seeker Company, 1894), p. 527.
40. Warren, *American Freethought,* pp. 161–62. The text of the Nine Demands appears on these pages.
41. Ibid., p. 162.
42. Ibid., p. 203.
43. Ibid., p. 160.
44. Flynn and Greeley, "Ingersoll, Robert Green," in Flynn, *The New Encyclopedia of Unbelief,* p. 426.
45. F. Forrester Church, *So Help Me God: The Founding Fathers and the First Great Battle over Church and State* (Orlando, FL: Harcourt, 2007), p. 233.
46. These charges are discussed in Dreisbach, *Thomas Jefferson and the Wall of Separation,* pp. 18–20.
47. James Turner, *Without God, Without Creed: The Origins of Unbelief in America* (Baltimore: Johns Hopkins University Press, 1985), p. 44.
48. Hamburger, *Separation of Church and State,* p. 121. The Republican in

question was the clergyman Tunis Wortman, writing in 1800 in "A Solemn Address to Christians and Patriots," ibid., p. 126.

49. Church, *The Separation of Church and State*, p. 18.
50. Maxine Seller, "Historical Perspectives on American Immigration Policy: Case Studies and Current Implications," *Law and Contemporary Problems* 45 (1982), p. 147.
51. Hamburger, *Separation of Church and State*, p. 202.
52. Ibid., p. 219.
53. Vincent Lannie and Bernard Diethorn, "For the Honor of Glory and God: The Philadelphia Bible Riots of 1840," *History of Education Quarterly* 8 (1968), p. 49.
54. Aside from including books that most Protestants do not see as canonical, and placing some books in a different order, the latter also provided annotations consisting of Catholic teachings on the scriptures.
55. Lannie and Diethorn, "For the Honor of Glory and God," p. 57.
56. Ibid., p. 71.
57. On this see Feldberg, *The Philadelphia Riots of 1844;* Ray Allen Billington, *The Protestant Crusade, 1800–1860: A Study of the Origins of American Nativism* (New York: Rinehart and Company, 1952); Diane Ravitch, *The Great School Wars, New York City, 1805–1973: A History of the Public Schools as Battlefield of Social Change* (New York: Basic Books, 1974). On the violence itself, see Elizabeth Geffen, "Violence in Philadelphia in the 1840s and 1850s," *Pennsylvania History* 36 (1969), pp. 381–410.
58. Lannie and Diethorn, "For the Honor of Glory and God," pp. 87, 94.
59. Feldman, *Divided by God*, p. 85. This position was more prominent in New York than in Philadelphia. In the latter nativists made arguments based on religious themes, not separationist ones. Also see Tracy Fessenden, "The Nineteenth-Century Bible Wars and the Separation of Church and State," *Church History* 74 (2005), p. 794.
60. Francis Graham Lee, *Church-State Relations* (Westport, CT: Greenwood Press, 2002), p. 48.
61. See Fessenden, "The Nineteenth-Century Bible Wars," pp. 784–811.
62. J. Scott Slater, "Florida's 'Blaine Amendment' and Its Effect on Educational Opportunities," *Stetson Law Review* 33 (2004), p. 588.
63. Larson, *American Infidel*, p. 119. Ingersoll went all in for Blaine

amendments, enthusing, "I insist that no religion should be taught in any school supported by public money; and by religion I mean superstition." Robert G. Ingersoll, "On Separation of Church and State," in *What's God Got to Do with It?: Robert Ingersoll on Free Thought, Honest Talk, and the Separation of Church and State,* edited by Tim Page (Hanover, NH: Steerforth Press, 2005), p. 116. Flynn and Greeley, "Ingersoll, Robert Green," in Flynn, *The New Encyclopedia of Unbelief,* p. 426. On Ingersoll's friendship with Blaine, see Larson, *American Infidel,* p. 119.

64. Notice that the issue was not federal funding but state funding, a concern that takes us back to the "incorporation" controversy discussed in Chapter 5.
65. Alfred W. Meyer, "The Blaine Amendment and the Bill of Rights," *Harvard Law Review* 64 (April 1951), p. 941.
66. Feldman, *Divided by God,* p. 86.
67. Slater, "Florida's 'Blaine Amendment,'" p. 584.
68. The original speech took place on September 12, 1960, and was addressed to the Greater Houston Ministerial Association, a Protestant organization. "Transcript: JFK's Speech on His Religion," *NPR* (5 December 2007), http://www.npr.org/templates/story/story.php?storyId=16920600.
69. Martin Luther King Jr., "A Knock at Midnight," in *Strength to Love* (Philadelphia: Fortress Press, 1963), p. 62.
70. Maurice Isserman and Michael Kazin, *America Divided: The Civil War of the 1960s,* 3rd ed. (New York: Oxford University Press, 2008).
71. Richard Rubenstein, *After Auschwitz: Radical Theology and Contemporary Judaism* (Indianapolis: Bobbs-Merrill, 1966), p. 246.
72. Ibid., p. 205.
73. Ibid., pp. 205, 225.
74. See, for example, Charles Bent, *The Death of God Movement* (Westminster, MD: Paulist Press, 1967). Also see Thomas Altizer, *The Gospel of Christian Atheism* (Philadelphia: Westminster Press, 1966).
75. *Gaudium et spes,* Pastoral Constitution on the Church in the Modern World, in *The Documents of Vatican II,* edited by Walter Abbott (New York: Crossroad, 1989), p. 216.
76. Ibid., p. 218.

77. Ibid., pp. 219–20. For a brief study of this issue, see the excellent work by Michael Gallagher, *What Are They Saying About Unbelief?* (New York: Paulist Press, 1995).
78. Karl Rahner, *Theological Investigations. Vol. 3: The Theology of Spiritual Life,* translated by Karl H. Kruger and Boniface Kruger (Baltimore: Helicon Press, 1967), p. 390.
79. Michel LeLong, *Pour un dialogue avec les athées* (Paris: Les Éditions du Cerf, 1965), pp. 138–39.
80. Carol Giardina, *Freedom for Women: Forging the Women's Liberation Movement, 1953–1970* (Gainesville: University Press of Florida, 2010), p. 195.
81. Davida Foy Crabtree, "Women's Liberation and the Church," in *Women's Liberation and the Church: The New Demand for Freedom in the Life of the Christian Church,* edited by Sarah Bentley Doely (New York: Association Press, 1970), p. 19.
82. Ibid., appendix, p. 99.
83. Nathan Abrams, "Triple-exthnics," *Jewish Quarterly* 196 (Winter 2004), http://www.jewishquarterly.org/issuearchive/articled325.html?articleid=38z.

7. *The Fall of American Secularism*

Epigraph: Francis Schaeffer, *A Christian Manifesto* (Wheaton, IL: Crossway Books, 1981).

1. Mark O'Keefe, "Robertson's Phone Corps Boosted GOP: Local Democrats Claim Network Ambushed Them," *The Virginian-Pilot* (9 November 1991), p. A1.
2. Reed's quarter-century friendship with the lobbyist Jack Abramoff threatened to destroy his political career after the scandal broke in the mid-2000s. In 2004, Reed admitted that "his public relations and lobbying companies had received at least $4.2 million from Abramoff to mobilize Christian voters to fight Indian casinos competing with Abramoff's casino clients." Thomas B. Edsall, "In Ga., Abramoff Scandal Threatens a Political Ascendancy," *Washington Post* (16 January 2006), http://www.washingtonpost.com/wp-dyn/content/article/2006/01/15/AR2006011500915.html. See also James Carney, "The Rise and

Fall of Ralph Reed," *Time* (23 July 2006), http://www.time.com/time/magazine/article/0,9171,1218060-1,00.html. Associated Press, "Does Christian Coalition Have a Prayer in '08?" *MSNBC.com* (3 March 2007), http://www.msnbc.msn.com/id/10958656/ns/politics/t/does-christian-coalition-have-prayer/.

3. Alex Marshall, "Religious Right Hoping to Influence Congressional Races: The Christian Coalition and Other Groups Are Trying to Repeat the Success They Had in the 1991 State Elections," *The Virginian-Pilot* (24 October 1992), p. A8.
4. For a biographical sketch, see Colin Duriez, "Francis Schaeffer," in *Handbook of Evangelical Theologians,* edited by Walter A. Fewell (Grand Rapids, MI: Baker Books, 1993), pp. 245–59.
5. As Schaeffer's son, Frank, has said, "And without the philosophical groundwork laid by these books and *A Christian Manifesto* – for which I served as Francis Schaeffer's research assistant – it is highly unlikely that people such as Pat Robertson, Jerry Falwell, James Dobson, Tim LaHaye, and others would have had the political influence they wield. This despite the fact that much of what comes out of the mouths of these people would today alarm Francis Schaeffer." John J. Whitehead, "Crazy for God: An Interview with Frank Schaeffer," *oldSpeak: An Online Journal Devoted to Intellectual Freedom,* Rutherford Institute (2010), http://www.rutherford.org/oldspeak/Articles/Interviews/oldspeak-frankschaeffer.html.
6. Schaeffer, *A Christian Manifesto,* p. 121. Though, to his credit, Schaeffer insisted that he was not endorsing theocracy.
7. Ibid., p. 36. Also see Barry Hankins, *Francis Schaeffer and the Shaping of Evangelical America* (Grand Rapids, MI: Eerdmans, 2008), p. 198.
8. Schaeffer, *A Christian Manifesto,* p. 54.
9. Ibid., p. 110. Although not specified in his book, Schaeffer seems to be railing against *McLean v. Arkansas Board of Education,* 529 F. Supp. 255 (Eighth Circuit 1983), which was then under way.
10. Francis Schaeffer, *How Should We Then Live?: The Rise and Decline of Western Thought and Culture* (Old Tappan, NJ: Fleming H. Revell Co., 1976), p. 226.
11. Cited in Goodman and Price, *Jerry Falwell,* p. 43. This was excerpted from the undated Moral Majority publication "Christians in Government: What the Bible Says."

12. See Berlinerblau, *Thumpin' It*, p. 145, n38. Randall Balmer, *Thy Kingdom Come: How the Religious Right Distorts the Faith and Threatens America: An Evangelical's Lament* (New York: Perseus, 2006), p. 12. And for the issue that did concern them, see Dirk Smillie, *Falwell Inc.: Inside a Religious, Political, Educational, and Business Empire* (New York: St. Martin's Press, 2008), p. 99.
13. Carter gained nearly half of the evangelical vote in 1976, up significantly from George McGovern's tallies in 1972. Albert Menendez, *Evangelicals at the Ballot Box* (Amherst, NY: Prometheus Books, 1996), p. 128. McGovern and the 1960s notwithstanding, evangelicals traditionally voted Democratic from the FDR period forward. Andrew Kohut, *The Diminishing Divide: Religion's Changing Role in American Politics* (Washington, DC: Brookings Institution, 2000).
14. Frederick Lane, *The Court and the Cross: The Religious Right's Crusade to Reshape the Supreme Court* (Boston: Beacon Press, 2008), p. 17. On Catholics being the first to respond to separationist decisions in the 1960s, see Joseph Preville, "Leo Pfeffer and the American Church-State Debate: A Confrontation with Catholicism," *Journal of Church and State* 33 (1991), p. 42.
15. As reported in Smillie, *Falwell Inc.*, p. 181.
16. Dahlia Lithwick, "Justice's Holy Hires," *Washington Post* (8 April 2007), http://www.washingtonpost.com/wp-dyn/content/article/2007/04/06/AR2007040601799.html. Also see Charlie Savage, "Scandal Puts Spotlight on Christian Law School," *Boston.com* (8 April 2007), http://www.boston.com/news/education/higher/articles/2007/04/08/scandal_puts_spotlight_on_christian_law_school/.
17. Kelly Boggs, "College Launches Pressler Law School," *Baptist Press* (22 August 2007), http://www.bpnews.net/bpnews.asp?id=26292. On another school devoted to the biblical worldview in Louisiana, see "Louisiana College to Create a Christian Law School," *Law.com* (27 August 2007), http://www.law.com/jsp/article.jsp?id=900005556959.
18. On this point, see George W. Dent Jr., "Secularism and the Supreme Court," *Brigham Young University Law Review* (1999), pp. 1–74.
19. *Arizona Sch. Choice Trust v. Winn*, 131 S. Ct. 1436, 1462 (2011).
20. *Flast v. Cohen*, 392 U.S. 83 (1968).
21. Linda Lyons, "The Gallup Brain: Prayer in Public Schools," *Gallup*

(10 December 2002), http://www.gallup.com/poll/7393/gallup-brain-prayer-public-schools.aspx.

22. Ibid.
23. Bill Mears, "Supreme Court Weighs Ten Commandments Cases," *CNN.com* (7 March 2005), http://edition.cnn.com/2005/LAW/03/02/scotus.ten.commandments/.
24. Linda Lyons, "Americans Indivisible on Pledge of Allegiance," *Gallup* (4 May 2004), http://www.gallup.com/poll/11551/Americans-Indivisible-Pledge-Allegiance.aspx.
25. Feldman, *Divided by God*, p. 163.
26. One scholar of Jewish involvement in the First Amendment made a point that for all intents and purposes could have been the war cry of the Revivalists who have set their sights on the court. Namely, all it takes is a new judicial precedent to "dash a group's collective hopes" as quickly as they were raised. Gregg Ivers, *To Build a Wall: American Jews and the Separation of Church and State* (Charlottesville: University Press of Virginia, 1995), p. 5.
27. Cited in Le Beau, *The Atheist*, p. 98.
28. Ronald Millar, "Strategy Trumps Precedent: Separationist Litigants on the Losing Side of Legal Change," *Journal of Church and State* 50 (2008), p. 303.
29. *Wallace v. Jaffree*, 472 U.S. 38 (1985).
30. Millar, "Strategy Trumps Precedent," p. 303.
31. Ibid., p. 313. Pfeffer left the American Jewish Congress in 1964 and served as special counsel for them after. He started PEARL in 1967. See Preville, "Leo Pfeffer," p. 46, n35. Eric Pace, "Leo Pfeffer, 83, Lawyer on Staff of the American Jewish Congress," *New York Times* (7 June 1993), http://www.nytimes.com/1993/06/07/obituaries/leo-pfeffer-83-lawyer-on-staff-of-the-american-jewish-congress.html.

    Some of the other Jewish groups that advocated for separation of church and state include the National Jewish Community Relations Advisory Council, the Synagogue Council of America, and the Jewish Committee on Law and Political Action. These smaller organizations have filed amici briefs with the Supreme Court in various landmark separationist cases, including *McCollum v. Illinois Board of Education*, 333 U.S. 203 (1948); *Engel v. Vitale*, 370 U.S. 421 (1962); *Abing-*

ton Township School District v. Schempp, 372 U.S. 203 (1963); and *Edwards v. Aguillard,* 482 U.S. 578 (1987). See the discussion in Leo Pfeffer, "Amici in Church-State Litigation," *Law and Contemporary Problems* 44 (1981), p. 86.

32. Ibid., p. 92. Also see Frank Sorauf, *The Wall of Separation: The Constitutional Politics of Church and State* (Princeton, NJ: Princeton University Press, 1976), pp. 52–53.
33. Russ Rankin, "Southern Baptists Decline in Baptisms, Membership, Attendance," *LifeWay* (9 June 2011), http://www.lifeway.com/Article/Southern-baptists-decline-in-baptisms-membership-attendance.
34. Millar, "Strategy Trumps Precedent," p. 320. See Aaron Douglas Weaver, "James M. Dunn and Soul Freedom: A Baptist Paradigm for Political Engagement in the Public Arena," *Baptist History and Heritage* (22 March 2010).
35. Pfeffer, "Amici," p. 92. Sorauf, *The Wall of Separation,* pp. 52–55.
36. Millar, "Strategy Trumps Precedent," p. 314. Sorauf, *The Wall of Separation,* p. 49. The American Federation of Teachers is an affiliate of the AFL-CIO. "About AFT: American Federation of Teachers, AFL-CIO," http://www.aft.org/about/. It has contributed amici briefs in a number of cases, including *Edwards v. Aguillard,* 482 U.S. 578 (1987). Likewise, the pro–public school National Education Association is the nation's largest union and has an active presence on the American political scene. It has also issued amici briefs in important church-state cases such as *Epperson v. Ark,* 393 U.S. 97 (1968) and *Committee for Public Education & Religious Liberty v. Nyquist,* 413 U.S. 756 (1973). The Horace Mann League has been involved in cases involving religion in schools [i.e., *Horace Mann League, Inc. v. Board of Public Works,* 242 Md. 645 (Md. Ct. App. 1966)] and states as one of its three primary purposes that of preserving "the separation of church and state and the use of government funds for only public schools." "Purpose of the HML," Horace Mann League of the USA, http://www.hmleague.org/index.php?option=com_content&view=article&id=1&Itemid=2.
37. Luther, "On Secular Authority," in Höpfl, *Luther and Calvin on Secular Authority,* p. 26.

38. Ibid., p. 25.
39. Locke, *A Letter Concerning Toleration,* p. 28 (emphasis mine).
40. Jefferson, "Notes on the State of Virginia," in Church, *The Separation of Church and State,* p. 51.
41. Jefferson, "Virginia Statute for Religious Freedom," in Church, *The Separation of Church and State,* p. 74.
42. Madison, "Memorial and Remonstrance," in Church, *The Separation of Church and State,* p. 61.
43. Ibid., p. 62.
44. Rousseau, *The Social Contract,* p. 73. See Witte, "The Essential Rights and Liberties," p. 7.
45. Jefferson, "Notes on the State of Virginia," in Church, *The Separation of Church and State,* p. 53.
46. Ed Dobson, "The Bible, Politics, and Democracy," in *The Bible, Politics, and Democracy,* edited by Richard Neuhaus (Grand Rapids, MI: Eerdmans, 1987), p. 3.
47. John Stuart Mill, "On Liberty," in *On Liberty and Other Essays,* edited by John Gray (New York: Oxford University Press, 2008),p. 18.
48. Ibid., pp. 49, 65.
49. Ibid., pp. 73, 77.
50. Ibid., p. 29.
51. Ibid., pp. 29, 37.
52. "If Israel were to keep two Sabbaths according to the laws thereof, they would be redeemed immediately." Isidore Epstein (ed.), *The Babylonian Talmud: Seder Mo'ed, Shabbath II* (London: Soncino, 1938), p. 582 (Shabbath 118b). "Umma," in *The Encyclopedia of Islam,* edited by Juan E. Campo (New York: Facts on File, Inc., 2009), pp. 687–88.
53. Sorauf, *The Wall of Separation,* p. 31.
54. Salter, "The Sordid Results of Humanism," in *The Humanist Threat,* p. 44. Also see Homer Duncan, *Secular Humanism: The Most Dangerous Religion in America* (Lubbock, TX: Christian Focus on Government, 1979), p. 7. How secular humanism came to be seen as a religion by the Supreme Court, no less, is a fascinating story told expertly by Leo Pfeffer, "The 'Religion' of Secular Humanism," *Journal of Church and State* 29 (1987), pp. 495–507.

8. Are Democrats Secularists?

1. "Obama's Speech on Faith-Based Organizations," *Real Clear Politics* (1 July 2008), http://www.realclearpolitics.com/articles/2008/07/obamas_speech_on_faithbased_or.html. Jeff Zeleny and Michael Luo, "Obama Seeks Bigger Role for Religious Groups," *New York Times* (2 July 2008), p. A1. Amy Sullivan, "Why Obama Seized the Faith-Based Mantle," *USA Today* (28 July 2008), p. 11A, http://www.usatoday.com/printedition/news/20080728/opledereligion136.art.htm.
2. One scholar writes, "Those who rejected the call for government intervention to uphold the cultural norms of particular religious and ethnic groups, coalesced around the Democratic Party. In the process they strengthened the salience of its identity as the party of secularism." Robin Archer, "American Communalism and Indian Secularism: Religion and Politics in India and the West," *Economic and Political Weekly* 34 (10–16 April 1999), p. 890.
3. The Pew Forum on Religion and Public Life, "Religious Belief and Public Morality: A Catholic Governor's Perspective; Governor Mario Cuomo, Remarks Delivered at the University of Notre Dame" (13 September 1984), Pew Research Center, http://pewforum.org/PublicationPage.aspx?id=611.
4. These arguments are developed in Berlinerblau, *Thumpin' It*.
5. See Jacques Berlinerblau, "Are Evangelicals Obama-Curious?" *On Faith* blog, *Washington Post* (25 February 2008), http://newsweek.washingtonpost.com/onfaith/georgetown/2008/02/areevangelicals_obamacurious.html.
6. An early observer of this problem was Stephen Carter, *The Culture of Disbelief: How American Law and Politics Trivialize Religious Devotion* (New York: Basic Books, 1993), p. 19.
7. Clinton's remark was transcribed from the author's notes of his coverage of the event. See Jacques Berlinerblau, "Religion and Politics Can Mix," *On Faith* blog, *Washington Post* (14 April 2008), http://newsweek.washingtonpost.com/onfaith/georgetown/2008/04/last_nights_compassion_forum.html.
8. FDCH E-Media, "Transcript: Illinois Senate Candidate Barack Obama,"

Washington Post (27 July 2004), http://www.washingtonpost.com/wp-dyn/articles/A19751-2004Jul27.html.

9. Obama, *The Audacity of Hope*, p. 39.
10. "Obama Announces White House Office of Faith-Based and Neighborhood Partnerships," White House Press Office (5 February 2009), http://www.whitehouse.gov/the_press_office/ObamaAnnouncesWhiteHouseOfficeofFaith-basedandNeighborhoodPartnerships/.
11. Sometimes this is referred to as nonpreferentialism. William Lester, "Student Religious Expression Within Public Schools," in Duncan and Jones, *Church-State Issues in America Today*, vol. 2, pp. 73–101. The terms are used as synonyms here, although for some scholars they refer to distinct legal approaches.
12. The Personal Responsibility and Work Opportunity Reconciliation Act of 1996, Pub. L. No. 104–193, 110 Stat. 2105 (1996).
13. Mary Segers, "President Bush's Faith-Based Initiative," in *Faith-Based Initiatives and the Bush Administration: The Good, the Bad, and the Ugly*, edited by Jo Renee Formicola, Mary Segers, and Paul Weber (Lanham, MD: Rowman and Littlefield, 2003), p. 6. Jeffrey Polet and David Ryden, "Past, Present, Future: Final Reflections on Faith-Based Programs," in *Sanctioning Religion?: Politics, Law, and Faith-Based Public Services*, edited by David Ryden and Jeffrey Polet (Boulder, CO: Lynne Reinner, 2005), p. 177.
14. Bob Wineburg, *Faith-Based Inefficiency: The Follies of Bush's Initiatives* (Westport, CT: Praeger, 2007), p. 84.
15. Segers, "President Bush's Faith-Based Initiative," in Formicola et al., *Faith-Based Initiatives*, p. 6.
16. Polet and Ryden, "Past, Present, Future," in Ryden and Polet, *Sanctioning Religion?*, p. 177.
17. George W. Bush, *A Charge to Keep* (New York: William Morrow, 1999), p. 213.
18. Ibid.
19. Ibid.
20. "Executive Order, Establishment of White House Office of Faith-Based and Community Initiatives," Office of the Press Secretary (29 January 2001), http://georgewbush-whitehouse.archives.gov/news/releases/2001/01/20010129-2.html (emphasis mine).
21. George Bush, "Remarks at the Office of Faith-Based and Commu-

nity Initiatives' National Conference" (26 June 2008), p. 917, http://www.gpo.gov/fdsys/pkg/WCPD-2008-06-30/pdf/WCPD-2008-06-30-Pg916.pdf.

22. See the articles of Sara Posner at *Religion Dispatches:* "Obama Under Fire from Civil Liberties Groups over Faith-Based Policies" (10 March 2010), http://www.religiondispatches.org/dispatches/sarahposner/2345/obama_under_fire_from_civil_liberties_groups_over_faith-based_policies; "The Right Attacks Obama's Religious Advisors" (25 August 2010), http://www.religiondispatches.org/dispatches/sarahposner/3212/the_right_attacks_obama%27s_religious_advisors; "Where's Obama's Faith-Based Office?" (9 September 2010), http://www.religiondispatches.org/dispatches/sarahposner/3305/where%27s_obama%27s_faith-based_office; "The ~~Bush~~ Obama Faith-Based Office" (3 December 2010), http://www.religiondispatches.org/dispatches/sarahposner/3835/the_bush_obama_faith-based_office/; "Obama's Faith-Based Office = The Kremlin?" (4 February 2011), http://www.religiondispatches.org/dispatches/sarahposner/4186/obama's_faith-based_office%3Dthe_kremlin.
23. John DiIulio Jr., *Godly Republic: A Centrist Blueprint for America's Faith-Based Future* (Berkeley: University of California Press, 2007), p. 141.
24. Wineburg, *Faith-Based Inefficiency,* p. 82. Adelle Banks, "Questions Linger on Faith-Based Makeover," *Christian Century* (10 March 2009), pp. 14–15. Adelle Banks, "Critics Say Action Needed at Faith-Based Office," *Christian Century* (2 November 2010), pp. 18–19.
25. Stanley Carlson-Thies, "Faith-Based Initiative 2.0: The Bush Faith-Based and Community Initiative," *Harvard Journal of Law and Public Policy* 32 (2009), p. 933.
26. Segers, "President Bush's Faith-Based Initiative," in Formicola et al., *Faith-Based Initiatives,* p. 7 (emphasis in original).
27. Urofsky, *Religious Freedom,* p. 164.
28. See Stephen Monsma, *Positive Neutrality: Letting Religious Freedom Ring* (Westport, CT: Greenwood Press, 1993), p. 194.
29. The Constitution of the United States, First Amendment.
30. Thomas Curry, *The First Freedoms: Church and State in America to the*

Passage of the First Amendment (New York: Oxford University Press, 1986), pp. 193, 216.

31. Levy, *The Establishment Clause,* pp. xix, xxi.
32. Henry Abraham, "Religion, the Constitution, the Court, and Society: Some Contemporary Reflections on Mandates, Words, Human Beings, and the Art of the Possible," in Goldwin and Kaufman, *How Does the Constitution Protect Religious Freedom?,* p. 22.
33. Drakeman, *Church-State Constitutional Issues,* p. 99.
34. Levy, *The Establishment Clause,* p. 5.
35. Stewart Davenport, "Dale's Laws," in Finkelman, *Religion and American Law,* pp. 119–20.
36. Curry, *The First Freedoms,* pp. 209–10.
37. Levy, *The Establishment Clause,* p. 26.
38. McClellan, "The Making and the Unmaking of the Establishment Clause," in McGuigan and Rader, *A Blueprint for Judicial Reform,* p. 295. Also see Robert Cord, *Separation of Church and State: Historical Fact and Current Fiction* (New York: Lambeth Press, 1982), pp. 5–6. David Ryden, "The Relevance of State Constitutions to Issues of Government and Religion," in Duncan and Jones, *Church-State Issues in America Today,* vol. 1, p. 232. Philip Kurland, "The Origins of the Religion Clause of the Constitution," *William and Mary Law Review* 27 (1986), pp. 839–61.
39. Levy, *The Establishment Clause,* pp. xxii, 76. Leonard notes, however, that none of these states had exclusive establishments.
40. Lester, "Student Religious Expression Within Public Schools," in Duncan and Jones, *Church-State Issues in America Today,* vol. 2, p. 76. John Witte Jr., *Religion and the American Constitutional Experiment* (Boulder, CO: Westview Press, 2005), pp. 156, 174. Levy, *The Establishment Clause,* p. 113.
41. *Wallace v. Jaffree,* 472 U.S. 38, 106 (1985).
42. On this point see Urofsky, *Religious Freedom,* p. 57.
43. See Witte, *Religion and the American Constitutional Experiment,* p. 154. For a discussion of how Washington came to exemplify this civic republican position, see Vincent Phillip Muñoz, "Religion and the Common Good: George Washington on Church and State," in Dreisbach et al., *The Founders on God and Government,* p. 10.

44. David Ryden and Jeffrey Polet, "Introduction: Faith-Based Initiatives in the Limelight," in Ryden and Polet, *Sanctioning Religion?*, p. 1.
45. For an overview of this issue, see Lewis Solomon, *In God We Trust?: Faith-Based Organizations and the Quest to Solve America's Social Ills* (Lanham, MD: Lexington Books, 2003).
46. Ibid., p. 2.
47. Douglas Koopman, "The Status of Faith-Based Initiatives in the Later Bush Administration," in Duncan and Jones, *Church-State Issues in America Today*, vol. 1, p. 174.
48. Ibid.
49. On this issue, see David Saperstein, "Public Accountability and Faith-Based Organizations: A Problem Best Avoided," *Harvard Law Review* 116 (2003), pp. 1353–96.
50. Polet and Ryden "Past, Present, Future," in *Sanctioning Religion?*, p. 188.
51. Explaining how Bush's executive order brought this into being is the former director himself: DiIulio, *Godly Republic*, p. 137.
52. Title VII of the Civil Rights Act of 1964, Pub. L. No. 88–352, 42 U.S.C. §2000e-1(a) states: "Inapplicability of subchapter to certain aliens and employees of religious entities: This subchapter shall not apply to an employer with respect to the employment of aliens outside any State, or to a religious corporation, association, educational institution, or society with respect to the employment of individuals of a particular religion to perform work connected with the carrying on by such corporation, association, educational institution, or society of its activities." Also see Nicholas Miller, "The Religious Right's Assault on the Rights of Religion: Undermining the Constitutionally Protected Status of Religious Schools," *Journal of Church and State* 42 (2000), p. 281.
53. Dana Milbank, "Charity Cites Bush Help in Fight Against Hiring Gays – Salvation Army Wants Exemption from Laws," *Washington Post* (10 July 2001), p. A1. Segers, "President Bush's Faith-Based Initiative," in Formicola et al., *Faith-Based Initiatives*, pp. 8–9, 134.
54. Ibid., p. 8.
55. Frank Bruni and Elizabeth Becker, "Charity Is Told It Must Abide by Antidiscrimination Laws," *New York Times* (11 July 2001), http://

www.nytimes.com/2001/07/11/us/charity-is-told-it-must-abide-by-antidiscrimination-laws.html.

56. See Koopman, "The Status of Faith-Based Initiatives," in Duncan and Jones, *Church-State Issues in America Today,* vol. 1, p. 170. Ryden and Polet, "Faith-Based Initiatives," in Ryden and Polet, *Sanctioning Religion?,* pp. 3–4.
57. "Obama Touts Faith-Based Plans," *USA Today* (2 July 2008), p. 5A.
58. Jane Lampman, "Obama's Vision of Faith-Based Programs," *Christian Science Monitor* 100 (3 July 2008). Zeleny and Luo, "Obama Seeks Bigger Role for Religious Groups."
59. Banks, "Questions Linger," p. 14. Adelle Banks, "Faith-Based Reform Gets Mixed Reviews," *Christian Century* (14 December 2010).
60. DiIulio, upon accepting the job, had agreed to work for only six months, thus becoming, in the words of Jo Renee Formicola and Mary Segers, "a lame duck before the shooting season even began." "The Ugly Politics of the Faith-Based Initiative," in Formicola et al., *Faith-Based Initiatives,* p. 128. Also see Amy Black and Douglas Koopman, "The Politics of Faith-Based Initiatives," in *Religion and the Bush Presidency,* edited by Mark J. Rozell and Gleaves Whitney (New York: Palgrave Macmillan, 2007), pp. 168–69.
61. Cited in Joseph Conn, "Religious Right, Politics Have Undue Influence at White House, Ex-Faith Czar DiIulio Charges," *Church and State* (January 2003), p. 7. His remarkable philippic can be found in "John DiIulio's Letter," *Esquire* (23 May 2007), http://www.esquire.com/features/dilulio.
62. "Towey Leaving Helm of Faith-Based Initiatives," *Christian Century* (16 May 2006), p. 14. Also see Wineburg, *Faith-Based Inefficiency,* p. xx.
63. All of the Bush office's directors were Roman Catholics or evangelicals. These were, in order, John DiIulio, Jim Towey, Jay Hein, and Jedd Medefind. More important, there seemed to be an almost willful preference for Christian groups in the office's outreach. David Kuo, author of the already mentioned tell-all, recounts a troubling conversation with a woman who reviewed applications from social-service agencies on behalf of the Bush administration. She shyly admits that she and her colleagues voted for grants only from Christian groups. Da-

vid Kuo, *Tempting Faith: An Inside Story of Political Seduction* (New York: Free Press, 2006), pp. 215–16.

64. "Faith Czar's Comments About 'Fringe Groups' Draw AU Criticism," *Church and State* (January 2004), p. 15.
65. Kuo, *Tempting Faith,* p. 141.
66. Ibid., pp. 229–30.
67. Also see Michael Fauntroy, "Buying Black Votes?: The GOP's Faith-Based Initiative," in Rozell and Whitney, *Religion and the Bush Presidency,* p. 182. And Formicola and Segers, "The Ugly Politics of the Faith-Based Initiative," in Formicola et al., *Faith-Based Initiatives,* p. 117.
68. Wineburg notes that Bush received 16 percent of the African American vote in 2004, up from 9 percent in 2000. *Faith-Based Inefficiency,* pp. xxiv, 88.
69. Kuo, *Tempting Faith,* pp. 206–7.
70. Ibid., p. 242.
71. Ibid., p. 262.
72. Banks, "Critics Say," p. 18.
73. Ibid. "Obama Announces White House Office of Faith-Based and Neighborhood Partnerships," Office of the Press Secretary (5 February 2009), http://www.whitehouse.gov/the_press_office/ObamaAnnouncesWhiteHouseOfficeofFaith-basedandNeighborhoodPartnerships/. Barack Obama, "Amendments to Executive Order 13199 and Establishment of the President's Advisory Council for Faith-Based and Neighborhood Partnerships," Office of the Press Secretary (5 February 2009), http://www.whitehouse.gov/the-press-office/amendments-executive-order-13199-and-establishment-presidents-advisory-council-faith.
74. *Walz v. Tax Commission,* 397 U.S. 664, 670 (1970). The full quotation reads, "No perfect or absolute separation is really possible; the very existence of the Religion Clauses is an involvement of sorts – one that seeks to mark boundaries to avoid excessive entanglement."
75. *Lemon v. Kurtzman,* 403 U.S. 602, 612 (1971). The "Lemon test" stipulates, "First, the statute must have a secular legislative purpose; second, its principal or primary effect must be one that neither advances nor inhibits religion . . . finally, the statute must not foster 'an excessive government entanglement with religion.'" Also see

Kenneth Ripple, "The Entanglement Test of the Religion Clauses: A Ten Year Assessment," *UCLA Law Review* 27 (1979–1980), p. 1197.

76. *Lemon v. Kurtzman,* 403 U.S. 602, 612 (1971).
77. "Faith-Based Programs Still Popular," Pew Research Center Publications (16 November 2009), http://pewresearch.org/pubs/1412/faith-based-programs-popular-church-state-concerns. Respondents did, however, express serious qualms about discriminatory hiring and trespasses of separation between church and state.

9. The Christian Nation and the GOP

1. "Overview," WallBuilders, LLC, http://www.wallbuilders.com/ABTOverview.asp.
2. John Fea, *Was America Founded as a Christian Nation?: A Historical Introduction* (Louisville, KY: Westminister John Knox, 2011), p. xiv. Kurt W. Peterson, "American Idol: David Barton's Dream of a Christian Nation," *Christian Century* (31 October 2006), p. 20.
3. See the description in Fea, *Was America Founded as a Christian Nation?*, pp. xiii–xiv, 57–59.
4. Peterson, "American Idol," p. 21.
5. Erik Eckholm, "Using History to Mold Ideas on the Right," *New York Times* (4 May 2011), http://www.nytimes.com/2011/05/05/us/politics/05barton.html.
6. Ibid.
7. See, for example, Nehemiah 2:11–18, 6:1.
8. "Overview," WallBuilders, LLC.
9. Eckholm, "Using History."
10. "David Barton: Propaganda Masquerading as History," A Report by People for the American Way Foundation (2011), http://aara.pfaw.org/media-center/publications/david-barton-propaganda-masquerading-history.
11. On this point see Isaac Kramnick and R. Laurence Moore, "The Godless Constitution," in *Protestantism and the American Founding,* edited by Thomas Engeman and Michael Zuckert (Notre Dame, IN: University of Notre Dame Press, 2004), pp. 129–42.

12. "The Treaty of Tripoli, Article 11," in Church, *The Separation of Church and State,* p. 123.
13. "An Ordinance for the government of the Territory of the United States northwest of the River Ohio," section 3 (1787), http://www.ourdocuments.gov/doc.php?flash=true&doc=8&page=transcript.
14. The Unanimous Declaration of the Thirteen United States of America (1776).
15. On these issues in general, see Jon Butler, "Why Revolutionary America 'Wasn't a Christian Nation,'" in Hutson, *Religion and the New Republic,* pp. 187–202. A diametrically opposed view is offered by Hutson, *Forgotten Features of the Founding,* pp. 111–32. Also, Fea's *Was America Founded as a Christian Nation?* scrutinizes the issue intelligently.
16. "Farewell Address" (selections) in Church, *The Separation of Church and State,* p. 118. For a sober and concise treatment of Washington's religiosity, see David Holmes, *The Faith of the Founding Fathers* (New York: Oxford University Press, 2006), pp. 59–71.
17. James Hutson (ed.), *The Founders on Religion: A Book of Quotations* (Princeton, NJ: Princeton University Press, 2005), p. 60. See Fea, *Was America Founded as a Christian Nation?,* pp. 233–37.
18. Hutson, *The Founders on Religion,* p. 59.
19. Fea, *Was America Founded as a Christian Nation?,* p. 21.
20. Thomas Curry, *The First Freedoms,* p. 219; also see pp. 195–97. Also see W. Jason Wallace, "Public Expression of Faith by Political Leaders," in Duncan and Jones, *Church-State Issues in America Today,* Vol. 1, pp. 112–13.
21. McCain made these remarks in an interview with Beliefnet: Dan Gilgoff, "John McCain: Constitution Established a 'Christian Nation,'" *Beliefnet,* http://www.beliefnet.com/Video/News-and-Politics/John-McCain-2008/John-Mccain-Constitution-Established-A-Christian-Nation.aspx.
22. The history of McCain's squabbles with conservative Christians over, among other things, the John Tower confirmation hearings and the McCain-Feingold Act, is recounted in Berlinerblau, *Thumpin' It,* pp. 120–25.
23. Ibid., p. 122.
24. Joel Connelly and Ed Offley, "McCain and Bush Clash over Revs. Rob-

ertson, Falwell," *Seattle Post-Intelligencer* (29 February 2000), p. A1. The original address can be found at "Sen. John McCain Attacks Pat Robertson, Jerry Falwell, Republican Establishment as Harming GOP Ideals," *CNN.com* (28 February 2000), http://transcripts.cnn.com/TRANSCRIPTS/0002/28/se.01.html.

25. John McCain, *Worth the Fighting For: The Education of an American Maverick and the Heroes Who Inspired Him* (New York: Random House, 2003), p. 136.
26. Gilgoff, "John McCain: Constitution Established a 'Christian Nation.'"
27. Bruce Smith, "McCain Says He's Been Baptist for Years," *Washington Post* (6 September 2007), http://www.washingtonpost.com/wp-dyn/content/article/2007/09/16/AR2007091600864.html.
28. On this dynamic, see also Jacques Berlinerblau, "McCain's Move: Deplorable but Effective," *On Faith* blog, *Washington Post* (4 October 2007), http://newsweek.washingtonpost.com/onfaith/georgetown/2007/10/mccains_move_deplorable_but_pr.html.
29. See, for example, Alan Dershowitz, "McCain and the Godless Constitution," *HuffingtonPost.com* (3 October 2007), http://www.huffingtonpost.com/alan-dershowitz/mccain-and-the-godless-co_b_66964.html.
30. Jon Meacham, "A Nation of Christians Is Not a Christian Nation," *New York Times* (7 October 2007), http://www.nytimes.com/2007/10/07/opinion/07meacham.html.
31. Tom Blackburn, "Christian Nation? Not Now, Not Ever," *Kansas City Star* (11 October 2007).
32. Jeremy Leaming, "Rumblings on the Right: Disgruntled Dobson Threatens Third Party at Secretive CNP Meeting," *The Wall of Separation* blog, Americans United for Separation of Church and State (1 October 2007), http://blog.au.org/2007/10/01/rumblings-on-the-right-disgruntled-dobson-threatens-third-party-at-secretive-cnp-meeting/.
33. Jacques Berlinerblau, "Couples Counseling for McCain and Evangelicals," *On Faith* blog, *Washington Post* (25 September 2007), http://newsweek.washingtonpost.com/onfaith/georgetown/2007/09/couples_counseling_for_mccain.html.
34. Her greatest collision with secularists on this issue came a few years after. Teddy Davis and Matt Loffmann, "Sarah Palin's 'Christian

Nation' Remarks Spark Debate," *ABCNews.com* (20 April 2010), http://abcnews.go.com/Politics/sarah-palin-sparks-church-state-separation-debate/story?id=10419289.

35. "'07 Survey Shows Americans' Views Mixed on Basic Freedoms," *First-amendmentcenter.org*.
36. Ibid.
37. "A New Century: A New Reformation," Forum Transcript, Washington National Cathedral (27 January 2008), http://www.nationalcathedral.org/learn/forumTexts/SF080127T.shtml.
38. "Transcripts: Saddleback Presidential Candidates Forum," *CNN.com* (17 August 2008), http://transcripts.cnn.com/TRANSCRIPTS/0808/17/se.01.html.
39. Jacques Berlinerblau, "McCain Wins with Home Field Advantage," *On Faith* blog, *Washington Post* (17 August 2008), http://newsweek.washingtonpost.com/onfaith/georgetown/2008/08/mccain_won_american_secularism.html.
40. He had previously made the same statement in an interview before the Saddleback summit meeting with McCain and Obama. Jeffrey Goldberg, "The Rick Warren Interview: No Compromise with Evil," *The Atlantic* (15 August 2008), http://www.theatlantic.com/international/archive/2008/08/the-rick-warren-interview-no-compromise-with-evil/8708/.
41. Christopher Hitchens, "Three Questions About Rick Warren's Role in the Inauguration," *Slate* (19 December 2008), http://www.slate.com/id/2207148/.
42. Ibid.
43. The original quote attributed to Warren is, however, hard to tease out from these sources: Hitchens, "Three Questions"; Lynda Resnick, "Aspen Institute Ideas Festival Speakers Dinner—July 8, 2010," *Lynda's Blog* (9 July 2010), http://blog.lyndaresnick.com/2010/07/aspen-institute-ideas-festival-speakers-dinner-july-8-2010/; Joe Klein, "The Not-So-Right Reverend Rick," *Swampland* blog, *Time* (18 December 2008), http://swampland.time.com/2008/12/18/the-not-so-right-reverend-rick/; Lynda Resnick, "The Unwarranted Talk About Warren," *Huffingtonpost.com* (22 December 2008), http://www.huffingtonpost.com/lynda-resnick/the-unwarranted-talk-abou_b_153008.html.

44. Rick Warren, "A New Century: A New Reformation," speech (27 January 2008), http://www.nationalcathedral.org/events/SF080127.shtml.
45. Andy Birkey, "Was Pastor Campbell's Prayer on the Senate Floor Legal?," *Minnesota Independent* (16 March 2011), http://minnesotaindependent.com/79058/was-pastor-campbells-prayer-on-the-senate-floor-legal.
46. Mark Souder, "A Conservative Christian's View on Public Life," in *One Electorate Under God: A Dialogue on Religion and American Politics*, edited by E. J. Dionne Jr., Jean Bethke Elshtain, and Kayla Drogosz (Washington, DC: Brookings Institution Press, 2004), p. 21.
47. Carl Hulse, "Citing Affair, Republican Gives Up House Seat," *New York Times* (18 May 2010), http.//www.nytimes.com/2010/05/19/us/politics/19souder.html. Maureen Dowd, "All the Single Ladies," *New York Times* (18 May 2010), http://www.nytimes.com/2010/05/19/opinion/19dowd.html. Carol D. Leonnig and Mary Ann Akers, "Ind. Rep. Mark Souder to Resign After Affair with Staffer," *Washington Post* (18 May 2010), http://www.washingtonpost.com/wp-dyn/content/article/2010/05/18/AR2010051803290.html.
48. Jackson Baker, "Tennessee Right to Life Endorses Fincher in 8th Congressional District," *Memphis Flyer* (6 July 2010), http://www.memphisflyer.com/JacksonBaker/archives/2010/07/06/tennessee-right-to-lfe-endorses-fincher-in-8th-congressional-district.
49. Brian Beutler, "LA GOPer: November a Choice Between an Atheist Society and a Christian Nation," *Talking Points Memo: TPMDC* (25 August 2010), http://tpmdc.talkingpointsmemo.com/2010/08/la-goper-november-a-choice-between-an-athiest-society-and-a-christian-nation.php.
50. The White House, "Joint Press Availability with President Obama and President Gul of Turkey, Cankaya Palace, Ankara, Turkey," Office of the Press Secretary (6 April 2009), http://www.whitehouse.gov/the_press_office/Joint-Press-Availability-With-President-Obama-And-President-Gul-Of-Turkey/.
51. Chris Steller, "Bachmann: Obama 'Bowing Before the King of Saudi Arabia and Then Lying About It,'" *Minnesota Independent* (11 April 2009), http://minnesotaindependent.com/31856/bachmann-obama-bowing-lying.

52. H.R. 2104 – 110th Congress: Public Prayer Protection Act of 2007 (2007). Retrieved from the Library of Congress, Thomas: http://thomas.loc.gov/cgi-bin/query/z?c110:H.R.2104:.
53. Ibid., §2(1).
54. Ibid., §2(2).
55. Ibid., §2(3).
56. Proposed amendment to 28 U.S.C. Chapter 8, cited in H.R. 2104 §3(a).
57. Ibid.
58. H. Con. Res. 121 – 111th Congress: Encouraging the President to designate 2010 as "The National Year of the Bible" (2009). Retrieved from the Library of Congress, Thomas: http://thomas.loc.gov/cgi-bin/query/z?c111:H.CON.RES.121:. See John Fea's discussion of H.R. 888 in *Was America Founded as a Christian Nation?*, p. xiv.
59. H. Res. 397 – 111th Congress: Affirming the rich spiritual and religious history of our Nation's founding and subsequent history and expressing support for designation of the first week in May as "America's Spiritual Heritage Week" for the appreciation of and education on America's history of religious faith (2009). Retrieved from the Library of Congress, Thomas: http://thomas.loc.gov/cgi-bin/query/z?c111:H.RES.397:.
60. Congressman J. Randy Forbes of Virginia, who introduced the resolution, did speak of Judeo-Christian civilization in his address to the house. J. Randy Forbes, "Judeo Christian Nation," http://www.forbes.house.gov/judeochristiannation/.
61. Levy, *The Establishment Clause*, p. xvii (emphasis mine).
62. On the progress of accommodationism in the states, see David Ryden, "The Relevance of State Constitutions to Issues of Government and Religion," in Duncan and Jones, *Church-State Issues in America Today*, vol. 1, pp. 227–55.
63. Ibid., p. 228.

10. *Who* Could *Be a Secularist?*

Epigraph: Jeffrey Stout, "2007 Presidential Address: The Folly of Secularism," *Journal of the American Academy of Religion*, vol. 27, no. 3 (2008).

1. "President Obama's Remarks at National Prayer Breakfast," Transcript, *Belief* blog, *CNN.com* (3 February 2011), http://religion.blogs.cnn.com/2011/02/03/president-obamas-remarks-at-national-prayer-breakfast/.
2. Though again, we must be cautious about confusing atheists with secularists. Benjamin Beit-Hallahmi, "The Likely Atheists," *The Guardian* (11 April 2011), http://www.guardian.co.uk/commentisfree/belief/2011/apr/11/atheists-research-scientists. Phil Zuckerman's 2009 survey of atheists, agnostics, and "secularists" reveals that the individuals most likely to fall into those categories are men, Asian Americans, individuals under age thirty, holders of higher education degrees, and residents of the West Coast and Northeast. Zuckerman, "Atheism, Secularity, and Well-Being," pp. 952–53.
3. The 173 million figure is presented by Barry Kosmin and Ariela Keysar in their *American Religious Identification Survey* of 2008. They found that 76 percent of the 228 million adults in the United States self-identified as Christians. Barry A. Kosmin and Ariela Keysar, *American Religious Identification Survey (ARIS) 2008, Summary Report*, Trinity College (March 2009), p. 3, http://www.americanreligionsurvey-aris.org/reports/ARIS_Report_2008.pdf.
4. A point related succinctly by Emmet Kennedy, "The Tangled History of Secularism," *Modern Age* 42 (2000), pp. 31–37. Also see Charles Taylor, "Modes of Secularism," in Bhargava, *Secularism and Its Critics*, pp. 31–70.
5. David Carlin, "The Gay Movement and Aggressive Secularism," *America* (23 September 1995), p. 16.
6. Baptist Press, "Chapman Urges 'A New Vision, New Voices, New Victories,'" *MorrisChapman.com* (19 February 2002), http://www.morrischapman.com/article.asp?id=17.
7. The Pew Forum on Religion and Public Life, "Global Survey of Evangelical Protestant Leaders," Pew Research Center (22 June 2011), p. 13, http://pewforum.org/uploadedFiles/Topics/Religious_Affiliation/Christian/Evangelical_Protestant_Churches/Global%20Survey%20of%20Evan.%20Prot.%20Leaders.pdf.
8. Matthew 22:21, here using the more famous wording of the King James Version.

9. Romans 13:1 (NRSV).
10. Romans 13:2 (NRSV).
11. Isaac Backus, "Baptist Appeals for Religious Liberty," in Church, *The Separation of Church and State,* p. 19. On Baptists in general, see William Brackney, "Separation of Church and State," in *Historical Dictionary of the Baptists,* 2nd ed. (Lanham, MD: Scarecrow Press, 2009), pp. 515–16. Some Baptists today, however, opt for total separation. See, for example, Baptist Joint Committee for Religious Liberty, "Government Funding," http://www.bjconline.org/index.php?option=com_content&task=category§ionid=4&id=20&Itemid=107.
12. On these groups' participation, see Leo Pfeffer, *God, Caesar, and the Constitution: The Court as Referee of Church-State Confrontation* (Boston: Beacon Press, 1975), p. 7. A more cautious assessment of these groups' participation is offered in Hamburger, *Separation of Church and State,* pp. 92–93.
13. On this and some of the residual anti-Catholicism in Americans United, see Sorauf, *The Wall of Separation,* p. 34.
14. The entire sermon is preserved in Perry Deane Young, *God's Bullies: Native Reflections on Preachers and Politics* (New York: Holt, Rinehart and Winston, 1982), pp. 310–17. The quoted material is on pp. 313 and 317. Falwell himself discussed his change of heart in his book *Strength for the Journey: An Autobiography* (New York: Simon and Schuster, 1987), p. 290.
15. Dirk Smillie, *Falwell Inc.,* p. 10. It is likely more accurate to describe Falwell as a fundamentalist.
16. Young, *God's Bullies,* p. 310.
17. On the conversation with Nixon in question, see "Transcript of White House Tape 043-161: Richard Nixon and Billy Graham," Presidential Recordings Program, University of Virginia (21 February 1973), http://whitehousetapes.net/transcript/nixon/043-161. On Billy Graham's friendship with the Reverend Norman Vincent Peale and attendance at the 1960 Montreux conference on Kennedy's candidacy and the Catholic voting bloc, see Carol V. R. George, *God's Salesman: Norman Vincent Peale and the Power of Positive Thinking* (New York: Oxford University Press, 1993), pp. 195–215.
18. Sarah Pulliam Bailey, "Q & A: Billy Graham on Aging, Regrets, and

Evangelicals," Interview, *Christianity Today* (21 January 2011), http://www.christianitytoday.com/ct/2011/januaryweb-only/qabillygraham.html?start=2.

19. And many other American faiths as well, but for purposes of brevity this trinity will suffice.
20. The thought has been intimated elsewhere. See Cimino and Smith, "Secular Humanism and Atheism Beyond Progressive Secularism," pp. 415, 416.
21. Gary Dorrien, *The Making of American Liberal Theology: Imagining Progressive Religion, 1805–1900* (Louisville, KY: Westminster John Knox Press, 2001). See Dorrien's chapter "Unitarian Beginnings." On the reaction in liberal theology against the Calvinist view of humans as sinful, wicked, and depraved, see Arthur Cushman McGiffert Jr., "Protestant Liberalism," in *Liberal Theology: An Appraisal, Essays in Honor of Eugene William Lyman,* edited by David Roberts and Henry Van Dusen (New York: Charles Scribner's Sons, 1942), pp. 106–20.
22. F. Forrester Church, *God and Other Famous Liberals: Reclaiming the Politics of America* (New York: Simon and Schuster, 1991), pp. xxi–xxii.
23. Dorrien, *The Making of American Liberal Theology.* On liberal Protestantism's core tenets, see Jeffrey Hensley, "Liberal Protestantism," in Hillerbrand, *The Encyclopedia of Protestantism,* vol. 3, pp. 1086–90.
24. Dorrien, *The Making of American Liberal Theology.*
25. On Catholicism, ibid., pp. 395–450.
26. Hanna Siegel, "Christians Rip Glenn Beck over 'Social Justice' Slam," *ABC News* (12 March 2010), http://abcnews.go.com/WN/glenn-beck-social-justice-christians-rage-back-nazism/story?id=10085008.
27. Peter Hodgson, *Liberal Theology* (Minneapolis: Fortress Press, 2007), p. 19.
28. The Jewish Anti-Defamation League notes that the "Mainline Protestant (Lutheran, Methodist, Episcopalian, Presbyterian, United Church of Christ) position . . . is more sympathetic to Palestinian interests than to Israel." Eugene Korn, "Meeting the Challenge: Church Attitudes Towards the Israeli-Palestinian Conflict," Anti-Defamation League (October 2002), http://www.adl.org/interfaith/meeting_challenge.asp. See, for example, David Waters, "Presbyterian Report Condemns Israeli Policies in Palestinian Territories," *On Faith* blog, *Washington*

Post (11 June 2010), http://newsweek.washingtonpost.com/onfaith/undergod/2010/06/presbyterian_report_condemns_israeli_policies_in_palestinian_territories.html.

29. See McGiffert, "Protestant Liberalism," in Roberts and Van Dusen, *Liberal Theology*, pp. 119–20.
30. Hensley, "Liberal Protestantism," in Hillerbrand, *The Encyclopedia of Protestantism*, vol. 3, p. 1088.
31. In fact, Dorrien defines liberal theology itself as a distancing from the idea of external authorities: "it is the idea that Christian theology can be genuinely Christian without being based upon external authority." *The Making of American Liberal Theology*, vol. 1, p. xiii.
32. On this idea, see Lyman Van Law Cady, "The Liberal Attitude Toward Other Religions," in Roberts and Van Dusen, *Liberal Theology*, p. 150.
33. It is important to recall, however, that earlier versions of liberal Protestantism were often not enthusiastic about religious diversity. In fact, many leaders in the social gospel movement envisioned America as a Christian nation. On this point, see the valuable study of Fea, *Was America Founded as a Christian Nation?*, pp. 22–42.
34. "Member Communions and Denominations," National Council of Churches USA, http://www.ncccusa.org/members/.
35. On this point, see Fea, *Was America Founded as a Christian Nation?* On the other hand, the openness of Protestantism to secular thought can be seen in the success of a work like the theologian Harvey Cox's *Secular City* (1965).
36. Dorrien, *The Making of American Liberal Theology*, p. xxi.
37. Hensley, "Liberal Protestantism," in Hillerbrand, *The Encyclopedia of Protestantism*, vol. 3, p. 1089. On the Revivalists' advantage over these faiths, see Berger, "Secularism in Retreat," in Esposito and Tamimi, *Islam and Secularism in the Middle East*, p. 41 (pp. 38–51).
38. Robert Putnam and David Campbell, *American Grace: How Religion Divides and Unites Us* (New York: Simon Schuster, 2010), p. 148.
39. Personal communication with the author, July 28, 2011.
40. Erika B. Seamon, "The Shifting Boundaries of Religious Pluralism in America Through the Lens of Interfaith Marriage" (Ph.D. diss., Georgetown University, 2011), pp. 291, 336.
41. Milton Himmelfarb, "Church and State: How High a Wall?" *Commentary* (July 1966), p. 23.

42. Jonathan D. Sarna, "Church-State Dilemmas of American Jews," in *Jews in Unsecular America,* edited by Richard John Neuhaus (Grand Rapids, MI: Eerdmans, 1987), p. 57.
43. Ivers, *To Build a Wall,* p. 2.
44. On the well-concealed acrimony between the agencies, see Cohen, *Jews in Christian America,* pp. 192, 194, and Ivers, *To Build a Wall,* p. 20.
45. After having worked on the McCollum case Pfeffer experienced "an almost uninterrupted string of successful appearances before the court over the next twenty-five years." Ivers, *To Build a Wall,* p. 71. Ivers describes Pfeffer as "the organized Jewish community's preeminent scholar, advocate and . . . strategist on church-state relations" (p. 101).
46. *Engel v. Vitale,* 370 U.S. 421, 424 (1962).
47. Leo Pfeffer, *Church, State, and Freedom* (Boston: Beacon Press, 1953), p. 605.
48. Associated Press, "Muslims Seek Probe of Mayor's 'Christian Community' Remarks," *USA Today* (8 February 2010), http://www.usatoday.com/news/religion/2010-02-08-christian-muslim_N.htm.
49. "CAIR: Who We Are: The Council on American-Islamic Relations as Defined by Its Actions and Statements over 16 Years of Community Service," Council on American-Islamic Relations (June 2010), http://www.cair.com/CivilRights/CAIRWhoWeAre.aspx.
50. "HAF Policy Brief: Prayer in Public Schools," Hindu American Foundation, Inc. (2004), http://www.hafsite.org/?q=media/pr/prayer-public-schools.
51. Mirin Kaur Phool, "Sikh American Legal Defense and Education Fund Radio Address Statement," Sikh American Legal Defense and Education Fund (SALDEF) (10 August 2005), http://rac.org/_kd/Items/actions.cfm?action=Show&item_id=1197&destination=ShowItem.
52. Central Tibetan Administration, "His Holiness the Dalai Lama Said Secularism Is the Basis of All Religions," Official Website of the Central Tibetan Administration (13 November 2006), http://www.tibet.net/en/print.php?id=1000&articletype=flashold.
53. FDCH E-Media, "Transcript: Illinois Senate Candidate Barack Obama," *Washington Post* (27 July 2004), http://www.washingtonpost.com/wp-dyn/articles/A19751-2004Jul27.html.
54. Libertarian Party, "Libertarian Party 2010 Platform," §1.1 (St. Louis: May 2010), http://www.lp.org/platform.

55. David Boaz, *Libertarianism: A Primer* (New York: Free Press, 1997), p. 109.
56. Jeffrey A. Miron, *Libertarianism from A to Z* (New York: Basic Books, 2010), p. 148.
57. David Kirby and David Boaz, "The Libertarian Vote in the Age of Obama," *Policy Analysis* 658 (21 January 2010) p. 1, http://www.cato.org/pubs/pas/pa658.pdf.
58. Libertarians themselves have recently noted that they are "increasingly a swing vote." Ibid., p. 3.
59. Analyzing a Harvard study, libertarian scholars noted that "secular centrists may be the best proxy for libertarian beliefs." Kirby and Boaz, "The Libertarian Vote," p. 10. The study in question is "The Political Personality of America's College Students: A Poll by Harvard's Institute for Politics" (March 2004), http://www.iop.harvard.edu/Research-Publications/Polling/Spring-2004-Youth-Survey/Executive-Summary.
60. Michael Barbaro, "Behind N.Y. Gay Marriage, an Unlikely Mix of Forces," *New York Times* (25 June 2011), http://www.nytimes.com/2011/06/26/nyregion/the-road-to-gay-marriage-in-new-york.html.
61. "Faith and Choice," Fact Sheets, NARAL Pro-Choice America Foundation, http://www.prochoiceamerica.org/media/fact-sheets/general-faith-choice.pdf.

11. How to Be Secularish (In Praise of "Secular Jews" and "Cafeteria Catholics")

Epigraph: Philip Roth, *Zuckerman Unbound* (New York: Vintage, 1981), p. 195.

1. See James Ault Jr., "Secular Humanism," in Wuthnow, *Encyclopedia of Politics and Religion*, 2nd ed., pp. 796–97. Also see John Swomley, *Religious Liberty and the Secular State: The Constitutional Context* (Amherst: Prometheus Books, 1987), pp. 118–40.
2. Schaeffer, *A Christian Manifesto*, p. 112.
3. Hankins, *Francis Schaeffer*, p. 177.

4. Duncan, *Secular Humanism,* p. 7 (emphasis in original). See Randall Balmer, "Secular Humanism" in *Encyclopedia of Evangelicalism* (Louisville, KY: Westminster John Knox Press, 2002), p. 516.
5. David E. Rosenbaum, "Of 'Secular Humanism' and Its Slide into Law," *New York Times* (22 February 1985), p. A16.
6. *Torcaso v. Watkins,* 367 U.S. 488, 495, fn 11 (1961).
7. See Tom Flynn, "Against the Seductions of Misbelief," in Flynn, *The New Encyclopedia of Unbelief,* p. 17.
8. Paul Kurtz, "*Reason* Interview," in *In Defense of Secular Humanism* (Amherst, NY: Prometheus Books, 1983), p. 265. Also see Malise Ruthven, *Fundamentalism: The Search for Meaning* (Oxford: Oxford University Press, 2004), pp. 24–25.
9. How secular humanism came to be seen as a religion by no less an authority than the Supreme Court is a fascinating story told expertly by Leo Pfeffer, "The 'Religion' of Secular Humanism," *Journal of Church and State* 29 (1987), p. 500. See Rosenbaum, "Of 'Secular Humanism' and Its Slide into Law."
10. On the small size of secular humanism, see Eldon Eisenach, "Secular Humanism," in *Encyclopedia of American Religion and Politics,* edited by Paul Djupe and Laura Olson (New York: Facts on File, 2003), pp. 408–9.
11. Ranjit Sandhu and Matt Cravatta (eds.), *Media-Graphy: A Bibliography of the Works of Paul Kurtz, Fifty-one Years: 1952–2003* (Amherst, NY: Center for Inquiry Translational, 2004). Featured here is also a list of all the organizations he founded.
12. Bill Cooke, "Paul Kurtz," in Joshi, *Icons of Unbelief,* p. 195.
13. "Humanist Manifesto I," in *Humanist Manifestos I and II,* edited by Paul Kurtz (Amherst, NY: Prometheus Books, 1973), p. 7.
14. "Humanist Manifesto II," in Kurtz, *Humanist Manifestos I and II,* p. 14.
15. Paul Kurtz et al., *A Secular Humanist Declaration,* p. 24.
16. "Humanism and Its Aspirations: Humanist Manifesto III, a Successor to the Humanist Manifesto of 1933," American Humanist Association (2003), http://www.americanhumanist.org/Who_We_Are/About_Humanism/Humanist_Manifesto_III.
17. Paul Kurtz, "Humanist Manifesto 2000: A Call for a New Plane-

tary Humanism," Art. IX, The Need for New Planetary Institutions, http://www.secularhumanism.org/index.php?section=main&page=manifesto.

18. Paul Kurtz, *What Is Secular Humanism?* (Amherst, NY: Prometheus Books, 2007), p. 16.
19. On the "blinking incomprehension" this term created, see Cooke, "Paul Kurtz," in Joshi, *Icons of Unbelief*, p. 204.
20. Rosenbaum, "Of 'Secular Humanism' and Its Slide into Law."
21. Mark Oppenheimer, "Closer Look at Rift Between Humanists Reveals Deeper Divisions," *New York Times* (1 October 2010), http://www.nytimes.com/2010/10/02/us/02beliefs.html.
22. "Humanist Manifesto II," in Kurtz, *Humanist Manifestos I and II*, pp. 23–24.
23. There have been other attempts unlike our own to rethink secularism. See, for example, the short but interesting article of Rick Salutin, "The New Secularism: Cultivating a Sober Tone of Doubt," *Canadian Dimension* 40 (2006), pp. 17–23.
24. Abdullahi An-Nai'm, *Islam and the Secular State: Negotiating the Future of Shari'a* (Cambridge, MA: Harvard University Press, 2010).
25. Jürgen Habermas, "Religion in the Public Sphere: Cognitive Presuppositions for the 'Public Use of Reason' by Religious and Secular Citizens," in *Between Naturalism and Religion: Philosophical Essays*, translated by Ciaran Cronin (Cambridge: Polity Press, 2008), p. 143.
26. M. Holmes Hartshorne, *The Faith to Doubt: A Protestant Response to Criticisms of Religion* (Englewood Cliffs, NJ: Prentice Hall, 1963), p. 100.
27. Cynthia Ozick, "Bloodshed," in *Bloodshed and Three Novellas* (New York: Alfred A. Knopf, 1976), p. 72.
28. Augustine, "The Spirit and the Letter," in *Augustine: Later Works*, vol. 8, translated by John Burnaby (Philadelphia: Westminster Press, 1955), p. 195.
29. This brings to mind the aside of D. L. Munby in his work *The Idea of a Secular Society and Its Significance for Christians:* "The Christian faith reveals the inexhaustible patience of God with men, because God cares for the freedom of human choice" (London: Oxford University Press, 1963), pp. 76–77.

30. The most famous theorist of this problematic was Max Weber in "The Social Psychology of the World Religions" and "Religious Rejections of the World and Their Directions," in *From Max Weber: Essays in Sociology*, translated and edited by H. H. Gerth and C. Wright Mills (New York: Galaxy, 1958).
31. Ecclesiastes 5:17 as translated in *Hebrew-English Tanakh* (Philadelphia: Jewish Publication Society, 1999).
32. George Santayana, *Soliloquies in England, and Later Soliloquies* (New York: Charles Scribner's Sons, 1922), p. 97.
33. Georges Moustaki, "La Philosophie," on *Ballades en balade: Sagesses et chemins de fortune* (Paris: Polygram, 1989).
34. Lara Vapnyar, *Memoirs of a Muse* (New York: Pantheon, 2006), p. 24, 25.
35. On the misquote, see David E. Cortesi, "Dostoevsky Didn't Say It: Exploring a Widely Propagated Misattribution," *Infidels.org* (2000), http://www.infidels.org/library/modern/features/2000/cortesi1.html.
36. Michael Novak follows this line of analysis in *No One Sees God: The Dark Night of Atheists and Believers* (New York: Doubleday, 2008), p. 52. The book is otherwise quite sophisticated and fair-minded. See my review "What's Your Blick? God or Science?" *Washington Post Book World* (28 September 2008).
37. Holyoake, *The Principles of Secularism*, p. 11.
38. "Humanist Manifesto II," in Kurtz, *Humanist Manifestos I and II*, p. 21.
39. Kurtz, "Humanist Manifesto 2000."
40. For some scholars Augustine's *saeculum* "is a profoundly sinister thing . . . a penal existence, marked by the extremes of misery and suffering, by suicide, madness" and so forth. P.R.L. Brown, "Saint Augustine," in *Trends in Medieval Thought*, edited by Beryl Smalley (Oxford, UK: Basil Blackwell, 1965), p. 11. In Augustine's words, human existence is a "life of misery, a kind of hell on earth." Saint Augustine, *Concerning the City of God Against the Pagans*, book 22, chapter 22, p. 1068. Other specialists, however, have advanced less gloomy assessments.
41. Brown, "Saint Augustine," in Smalley, *Trends in Medieval Thought*, p. 11.
42. Robert Markus, *Christianity and the Secular* (Notre Dame, IN: Uni-

versity of Notre Dame Press, 2006), p. 10. In between Brown and Markus one thinks of Frederick Russell's view that Augustine was ambivalent about the *saeculum:* "'Only Something Good Can Be Evil': The Genesis of Augustine's Secular Ambivalence," *Theological Studies* 51 (1990) pp. 698–716.

43. Hannah Arendt, *Love and Saint Augustine* (Chicago: University of Chicago Press, 1996), p. 103.
44. Wilfred McClay, "Two Concepts of Secularism," in *Religion Returns to the Public Square: Faith and Policy in America,* edited by Hugh Heclo and Wilfred McClay (Baltimore: Johns Hopkins University Press, 2003), p. 53 (emphasis in original).
45. Elliot Dorff, *The Way into Tikkun Olam (Repairing the World)* (Woodstock, VT: Jewish Lights, 2005), p. 12.
46. This theme is explored at length in Berlinerblau, *The Secular Bible.*
47. Luther, "On Secular Authority," in Höpfl, *Luther and Calvin on Secular Authority,* p. 26.
48. Jefferson, "Notes on the State of Virginia," in Church, *The Separation of Church and State,* pp. 51–52.
49. Mill, "On Liberty," in *On Liberty and Other Essays,* pp. 16–17.
50. On the centrality of freedom of expression for Holyoake, see *The Principles of Secularism,* p. 15. In the same work Holyoake cites Mill (p. 35).
51. Roy, *Secularism Confronts Islam,* p. 38.
52. Ibid.
53. Malise Ruthven, *A Satanic Affair: Salman Rushdie and the Rage of Islam* (London: Chatto and Windus, 1990).
54. Andrew Anthony, "How One Book Ignited a Culture War," *The Guardian* (11 January 2009), http://www.guardian.co.uk/books/2009/jan/11/salman-rushdie-satanic-verses.
55. Some scholars have explored this concept through use of the term "the secularly religious." Berlinerblau, *The Secular Bible.* Also see Edward Bailey, "Secular Religion," in Swatos, *Encyclopedia of Religion and Society,* p. 456; A. L. Greil, "Secular Religions," in *International Encyclopedia of the Social and Behavioral Sciences,* vol. 20, edited by Neil Smelser and Paul Baltes (Amsterdam: Elsevier, 2001), pp. 13783–786; also see Clayton Crockett (ed.), *Secular Theology: American Radical*

Theological Thought (London: Routledge, 2001). John MacQuarrie, *God and Secularity* (Philadelphia: Westminster Press, 1979).

56. Tom Beaudoin, "Paul of Tarsus and Catholicism Today: A Response to Cardinal Rodriguez and Father Schreiter" (26 March 2009), National Pastoral Life Center, http://www.nplc.org/events/25_beaudoin.php. Of course, this view is not a popular one among Catholics. See, for example, the short entry "Secularism" in *The HarperCollins Encyclopedia of Catholicism,* edited by Richard McBrien (San Francisco: HarperCollins, 1989), p. 1180.
57. Ibid.
58. Ibid.
59. See Sidney Brichto, "Judaism in a Secular World," in *Two Cheers for Secularism,* edited by Sidney Brichto and Richard Harries (Northamptonshire, UK: Pilkington Press, 1998), pp. 23–35.
60. Secular Jews in the sense that we have already defined them (that is, focused solely on disestablishment without the *-ish* qualities) are rare indeed. Ultra-orthodox Jews in the United States, for example, tend not to support separationist positions. Among Jews, secular and secularish tend to blur together.
61. For some basic materials about this denomination, see Renee Kogel and Zev Katz (eds.), *Judaism in a Secular Age: An Anthology of Secular Humanistic Jewish Thought* (Farmington Hills, MI: KTAV Publishing House, 1995); Sherwin Wine, *Judaism Beyond God* (Farmington Hills, MI: KTAV Publishing House, 1995). The group has also produced a helpful manual by Eva Goldfinger titled "Basic Ideas of Secular Humanistic Judaism," published by the International Institute for Secular Humanistic Judaism in 1996.
62. The number rockets to 64 percent when the category becomes merely Jews who have Jewish parentage but say they have no religion. For this see Egon Mayer, Barry Kosmin, and Ariela Keysar, "Exhibit 9: Outlook of Jews by Religion and Adherents of Selected Other Religious Groups," in *American Jewish Identity Survey 2001* (New York: Center for Cultural Judaism, 2003), p. 35. Methodologically the study suffers from a flaw: the interviewers never defined the term *secular* for their respondents. Ergo, we do not quite know what those 44 percent thought secularism entailed. For a similar critique, see Zvi Gitelman,

"Conclusion: The Nature and Viability of Jewish Religious and Secular Identities," in *Religion or Ethnicity?: Jewish Identities in Evolution,* edited by Zvi Gitelman (New Brunswick, NJ: Rutgers University Press, 2009), pp. 303–22. A more recent paper on these findings by Ariela Keysar is "Secular Jews and Other Secular Americans: New Findings," paper presented at the Fifteenth World Congress of Jewish Studies (Jerusalem: August 2–6, 2009).

63. Mayer, Kosmin, and Keysar, *American Jewish Identity Survey 2001,* p. 35.
64. Uri Ram, "Why Secularism Fails: Secular Nationalism and Religious Revivalism in Israel," *International Journal of Politics, Culture and Society* 21 (2008), pp. 57–73.
65. For discussions about whether sustaining a life around a language is a feasible project, see the many contributions in Barnett Zumoff and Karl Zukerman (eds.), *Secular Jewishness for Our Time* (New York: The Forward Association, 2006), p. x.
66. Goodman makes this remark in a discussion of Chaim Zhitlowsky (1865–1943), one of the central figures in the Yiddish-based secular movement. Saul Goodman, "Introduction," in *The Faith of Secular Jews,* edited by Saul Goodman (New York: KTAV Publishing House, 1976), p. 10.
67. Barnett Zumoff and Karl Zukerman, "Editors' Foreword," in Zumoff and Zuckerman, *Secular Jewishness for Our Time,* p. x.
68. Yaakov Malkin (ed.), *Free Judaism and Religion in Israel* (Jerusalem: Free Judaism Quarterly, 1999), p. 87.

12. Tough Love for American Secularism

Epigraph: Agnès Poirier, "The Pope's Plot Against Secularism," *New Statesman* (18 September 2008).

1. Diane Cardwell, "Lieberman Talks Up Religion as Democrats Poke, Jostle, Warn, and Joke," *New York Times* (24 December 2003), http://www.nytimes.com/2003/12/24/politics/campaigns/24TRAI.html.
2. In a seminal essay Wilfred McClay asks, "Is it accurate to speak of sec-

ularism as a kind of substitute religion, a reservoir of ultimate beliefs about ultimate things that stands, in that sense, in a continuum with, and in competition with, conventional orthodox religious faiths?" McClay, "Two Concepts of Secularism," in Heclo and McClay, *Religion Returns to the Public Square*, p. 44.

3. Jacques Berlinerblau, "Iowa Faith and Freedom Coalition: Five Republicans Fight for the Faith(ful) in Iowa," *On Faith* blog, *Washington Post* (8 March 2011), http://onfaith.washingtonpost.com/onfaith/georgetown/2011/03/republicans_fight_for_god_at_the_iowa_faith_and_freedom.html.
4. Olson, *The Westminster Handbook to Evangelical Theology*, p. 6.
5. Jonathan Romain, "God, Doubt, and Dawkins," *European Judaism* 41 (2008), p. 71.
6. Ibid., p. 73. This quote is attributed to the rabbi Paul Freedman.
7. AU reports that it has "more than 120,000 members and supporters from all over the country." Americans United for Separation of Church and State, "Frequently Asked Questions About Americans United," http://www.au.org/about/faqs/.
8. Herbert David Rix, *Martin Luther: The Man and the Image* (Valley Stream, NY: Ardent Media, Inc., 1983), p. 85.
9. Nussbaum, *Liberty of Conscience*, pp. 273–81.
10. On this point see Witte, "That Serpentine Wall of Separation," p. 1900.
11. See "Syllabus of Errors" (extracts) in Maclear, *Church and State in the Modern Age*, p. 163.
12. Ibid., p. 166. This point about some of the irrational Catholic positions of the era is made in a critique of Philip Hamburger by John Witte Jr., "That Serpentine Wall of Separation," p. 1898.
13. Vatican II, for instance, declared, "This Vatican Council declares that the human person has a right to religious freedom. This freedom means that all men are to be immune from coercion on the part of individuals or of social groups and of any human power, in such wise that no one is to be forced to act in a manner contrary to his own beliefs, whether privately or publicly, whether alone or in association with others within due limits." Pope Paul VI, "Declaration on Religious Freedom, Dignitatis Humanae," Second Vatican Council (7 December 1965), http://www.cin.org/v2relfre.html.

14. See the discussion of this issue in Stephen Feldman, *Please Don't Wish Me a Merry Christmas: A Critical History of the Separation of Church and State* (New York: New York University Press, 1997), pp. 218–28.
15. Preville, "Leo Pfeffer," pp. 37–53.
16. Clarke Cochran and David Cochran, *The Catholic Vote: A Guide for the Perplexed* (Maryknoll, NY: Orbis, 2008), p. 37.
17. Ibid., pp. 34–36.
18. Lane, *The Court and the Cross,* p. 23.
19. That sort of tinkering will infuriate African Americans, proponents of immigration, those who don't want there to be fifty independent militias in the United States, women who would have their reproductive freedoms eradicated via plebiscite, and so many more. If Revivalists want to get into this fight, they are more than welcome to do so.
20. The Pew Forum on Religion and Public Life, "U.S. Religious Landscape Survey: Religious Affiliation: Diverse and Dynamic," Pew Research Center (February 2008), p. 12, http://religions.pewforum.org/pdf/report-religious-landscape-study-full.pdf.
21. On pervasive sectarianism, see Saperstein, "Public Accountability and Faith-Based Organizations," p. 1380.
22. A somewhat similar suggestion is offered by Guliuzza, *Over the Wall,* p. 148.
23. Vincent Philip Muñoz, "James Madison's Principle of Religious Liberty," *American Political Science Review* 97 (2003), p. 25.
24. Cited in Madan, "Secularism in Its Place," in Bhargava, *Secularism and Its Critics,* p. 311. Madan is citing Jawaharlal Nehru.
25. Segers, "President Bush's Faith-Based Initiative," in Formicola et al., *Faith-Based Initiatives,* pp. 1, 144.
26. Zev Chafets, "The Huckabee Factor," *New York Times Magazine* (12 December 2007), http://www.nytimes.com/2007/12/12/magazine/16huckabee.html?hp.
27. CBS/AP, "Huckabee Apologizes for Mormon Remark," *CBS News* (12 December 2007), http://www.cbsnews.com/stories/2007/12/12/politics/main3612914.shtml.
28. Brian Ross, "McCain Pastor: Islam Is a 'Conspiracy of Spiritual Evil,'" *ABC News* (22 May 2008), http://abcnews.go.com/Blotter/story?id=4905624&page=1.
29. Andy Sullivan, "Pro-Israel Evangelical Leader Endorses McCain,"

Reuters (27 February 2008), http://www.reuters.com/article/2008/02/28/us-usa-politics-mccain-idUSN2749859920080228.

30. Kimberly Kindy, "In Rebuking Minister, McCain May Have Alienated Evangelicals," *Washington Post* (29 May 2008), http://www.washingtonpost.com/wp-dyn/content/article/2008/05/28/AR2008052803037.html.
31. Joshua Green, "Michele Bachmann's Church Says the Pope Is the Antichrist," *The Atlantic* (13 July 2011), http://www.theatlantic.com/politics/archive/2011/07/michele-bachmanns-church-says-the-pope-is-the-antichrist/241909/. James Oliphant, "Michele Bachmann Leaves Church Accused of Anti-Catholic Bias," *Los Angeles Times* (5 July 2011), http://articles.latimes.com/2011/jul/15/news/la-pn-bachmann-church-20110715.
32. The White House, "Remarks by the President at Easter Prayer Breakfast," Office of the Press Secretary (19 April 2011), http://www.whitehouse.gov/the-press-office/2011/04/19/remarks-president-easter-prayer-breakfast. For discussion of this event see Jacques Berlinerblau, "Obama's Easter Christology: What Rankles, What Works," *Brainstorm* blog, *Chronicle of Higher Education* (20 April 2011), http://chronicle.com/blogs/brainstorm/obamas-easter-prayer-breakfast-christology-what-rankles-what-works/34400.
33. Chris Tomlinson, "Perry Invites Governors to Prayer Meeting," *Associated Press Archive* (6 June 2011). "Gov. Perry Declares August 6th a Day of Prayer," Proclamation, Office of the Governor Rick Perry (6 June 2011). "Why?" subsection of "Historic Response," The Response: A Call to Prayer for a Nation in Crisis, http://theresponseusa.com/why-the-response.php.
34. *Freedom from Religion Foundation, Inc. v. Obama,* 641 F.3d 803; 2011 U.S. App. (2011 Seventh Circuit).
35. J. Philip Wogaman, "The Churches and Legislative Advocacy," *American Academy of Political and Social Science* 446 (1979), p. 53.
36. Stout, "2007 Presidential Address: The Folly of Secularism," p. 535.
37. Chris Stedman, "Dear Religious Americans, How Many Atheists Do You Know?" *On Faith* blog, *Washington Post* (20 July 2011), http://www.washingtonpost.com/blogs/on-faith/post/dear-religious-americans-how-many-atheists-do-you-know/2011/07/20/gIQAfo8kPI_blog.html.

38. *Wallace v. Jaffree,* 472 U.S. 38, 106 (1985).
39. Concluding Document of the Vienna Meeting 1986 of Representatives of the Participating States of the Conference on Security and Co-operation in Europe, Held on the Basis of the Provisions of the Final Act Relating to the Follow-up to the Conference (Vienna, 1989), http://www.osce.org/mc/16262, Principle 16.1.
40. Elizabeth Shakman Hurd, *The Politics of Secularism in International Relations* (Princeton, NJ: Princeton University Press, 2008), p. 16. Also see William Connolly, *Why I Am Not a Secularist* (Minneapolis: University of Minnesota Press, 1999). Nandy, "The Politics of Secularism," in Bhargava, *Secularism and Its Critics,* p. 324. Jakobsen and Pellegrini, "Introduction: Times like These," in Jakobsen and Pellegrini, *Secularisms,* p. 7.
41. Smith, "India as a Secular State," in Bhargava, *Secularism and Its Critics,* p. 179.

Index